Space Industrialization

Volume I

Editor

Brian O'Leary, Ph.D.
Consultant
31272 Flying Cloud Drive
Laguna Niguel, California

CRC Press, Inc.
Boca Raton, Florida

Library of Congress Cataloging in Publication Data
Main entry under title:

Space industrialization.

   Bibliography: p.
   Includes index.
   1. Space stations—Industrial applications—
Addresses, essays, lectures. I. O'Leary,
Brian, 1940-
TL797.S64   629.44   80-26428
ISBN 0-8493-5890-6 (v. 1)
ISBN 0-8493-5891-4 (v. 2)

   Direct all inquiries to CRC Press, Inc., 2000 N.W. 24th Street, Boca Raton, Florida 33431.

© 1982 by CRC Press, Inc.

International Standard Book Number 0-8493-5890-6 (Volume I)
International Standard Book Number 0-8493-5891-4 (Volume II)

Library of Congress Card Number 80-26428
Printed in the United States

PREFACE

A few years ago, space industrialization was considered as a fanciful futuristic concept. Once it was a subject of speculation in narrow circles and bombarded by skepticism (including a fair degree of my own).

This is no longer true. The exponential growth of literature in the field, of which this volume comprises a significant step, is testimony to the seriousness of these ideas, their feasibility, and enormous potential for improving the human condition. The imminent onset of the space shuttle as an Earth-to-orbit workhorse assures the beginnings of space industrialization.

There are two types of space industrialization: (1) near-term, small products which can be produced by machinery which can be contained in one space shuttle cargo bay ("orbital jewelry", utilizing the uniqueness of a weightless and zero gravity environment, i.e., crystal growth, ball bearings, and pharmaceuticals) and (2) the processing of nonterrestrial materials into large space structures such as satellite power stations and space habitats. The former operations require payloads on the order of a few tons, while the latter involves the processing of millions of tons of nonterrestrial materials. Near-term programs will begin during the 1980s and will help to form an evolutionary foundation for a lunar or asteroidal materials processing enterprise that could begin as early as 10 years from now.

Political rather than technical considerations will probably determine the pace of development of space industrialization. Convincing economic cases can be made for a wide variety of facets of space industrialization, but, for the most part, it is still early for industry to take the necessary risks.

This will change, perhaps sooner than we now anticipate, as history has so often shown. The concepts described in this volume are likely to lead to revolutions in worldwide energy supply, communications, weather prediction, and health care. Toward the turn of the century it is conceivable that lunar and asteroidal materials processing could produce self-sufficient and permanent human habitats in space, ships of exploration of the solar system, a revolution in astronomy, and the potential for providing the world with abundant energy, food, and natural resources.

These volumes encompass both near-term and long-term programs of space industrialization. It is important to make clear distinctions between the two: near-term programs involve innovative products fabricated from materials launched from the gravity-well of the Earth, whereas the long-term programs involve the transport and processing of nonterrestrial materials. Both aspects are valid and far-reaching areas of inquiry.

The breadth of coverage can be gleaned by glancing at the Table of Contents. There are some areas which are elucidated in greater detail elsewhere and the interested reader is encouraged to key into the references given at the end of the chapters. Omitted are some aspects of innovative space system design (mass-drivers and mass-catchers, for example), trajectories, and many of the social, political, organizational, and human aspects of space industrialization. Breadth in the basic concepts rather than details is emphasized, with a multidisciplinary approach.

I appreciate the help of all contributing authors who took a careful and professional effort and gave generously of their valuable time. This book would not have been created without their contributions.

Brian O'Leary

THE EDITOR

Brian O'Leary, Ph.D., is a scientist, author, and former astronaut.

O'Leary is an authority on U.S. science policies and space exploration, and he has written more than 60 scientific papers, most of them dealing with planetary science. He has lectured to 100 college audiences and appeared on television programs advocating a bolder national effort in space.

His work has appeared in the *New York Times, Science, Omni, Quest 81, The Bulletin of Atomic Scientists, The New Scientist,* and other national publications. His book, *The Making of an Ex-Astronaut,* was awarded best young adult book of 1970 by the American Library Association. A second book, *The Fertile Stars,* will be published in summer 1981.

In 1967 and 1968, Dr. O'Leary was an astronaut at the Manned Spacecraft Center, Houston.

He has a B.A. in physics from Williams College, an M.A. in astronomy from Georgetown University, and a Ph.D. in astronomy from the University of California in Berkeley. He has been a member of faculties at Cornell University, California Technical Institute, the University of California at Berkeley, the University of Massachusetts, Princeton University, San Francisco State University, Hampshire College, University of Pennsylvania, and the University of Texas.

O'Leary espouses space as an energy and material resource for Earth. It is possible within the next 10 to 15 years, he believes, to establish space communities, satellite power stations, and manufacturing facilities which would make use of materials mined from the moon and the asteroids. He was the first to research the idea that the Earth approaching asteroids could provide cost-effective resources for space manufacturing.

CONTRIBUTORS

Lee Browning
Chief, Advanced Design
General Dynamics, Convair Division
San Diego, California

T. Stephen Cheston, Ph.D.
Acting Dean, Graduate School
Georgetown University
Washington, D.C.

David R. Criswell, Ph.D.
Visiting Research Physicist
California Space Institute
La Jolla, California

Hubert P. Davis, Ph.D.
Chief, Transportation Systems Office
Lyndon B. Johnson Space Center
Houston, Texas

Gerald W. Driggers
Senior Research and Development
    Engineer
Combustion Engineering, Inc.
Cropwell, Alabama

Peter E. Glaser, Ph.D.
Vice President
Arthur D. Little, Inc.
Cambridge, Massachusettes

Gerald K. O'Neill, Ph.D.
Institute for Space Studies
Princeton, New Jersey

J. Peter Vajk, Ph.D.
Senior Scientist
Science Applications, Inc.
Pleasanton, California

Robert D. Waldron, Ph.D.
Research Specialist
Space Systems Group
    Rockwell International
Downey, California

David L. Winter, M.D.
Director Medical Research
Sandoz, Inc.
East Hanover, New Jersey

TABLE OF CONTENTS

Volume I

Volume II

Chapter 1

SPACE INDUSTRIALIZATION: AN OVERVIEW

Gerald W. Driggers

TABLE OF CONTENTS

I. INTRODUCTION

Space Industrialization (SI) is the medium by which services, energy, and products are returned from space to Earth to provide economic and other pragmatic benefits to mankind. Although most studies to date have focused on the U.S. as the mechanism for benefit generation and transfer (with an appropriate payback to its industry and citizenry for investing resources and labor), it is the world that benefits. Indeed, the underdeveloped and developing countries are now, and will continue to be, prime beneficiaries from SI. It is possible to construct credible scenarios which step these nations into the twentieth century equivalent of the U.S. in less than 100 years, without significant local or global-economic or environmental damage. The great power for what is considered "good" in the western world (health, safety, knowledge, creative growth, etc.) afforded by SI has been comprehended by a very few, but there is evidence that realization is spreading.

Recent studies have examined the broad spectrum of SI and what may be gained from the investing of resources, both public and private, in the next decade.[1,2] The future was examined to characterize resource pressures, requirements, and supply (population, energy, materials, food); also, the backdrop of probable events, attitudes, and trends against which SI will envolve were postulated. The opportunities for space industry that would bring benefits to Earth were compiled and screened against terrestrial alternatives. Most survived, and a population of the survivors was examined to determine if SI would ever be "worth the investment". A cursory market survey was conducted for the selected services and products provided by these initiatives and the results were astounding. SI is a billion dollar a year business now; in 30 years it could grow by 100 times that amount or more!

Other studies have examined specific aspects of SI, such as production of energy via satellite,[3] use of nonterrestrial materials for manufacturing,[4] and processing of materials in space to manufacture high-value products.[5] These will be examined individually in later chapters. The following sections deal with the results of studies examining the interrelation of the various aspects of SI.

II. THE TERRESTRIAL FRAMEWORK FOR SI

During the next few decades, space technology (developed for purely scientific reasons, for political and prestige reasons, or to serve specific military needs) can be adapted, extended, and expanded to use the new environment and nearly limitless resources of outer space for the benefit of humanity in an economically profitable manner. SI will then grow from a handful of commerically operated communications satellites into a highly diversified and expanding sector of human socioeconomic system. In the first few decades, however, it will necessarily depend for its very existence on the conventional segments of the socioeconomic system to provide the technology, the original investment capital, and the markets for its goods and services. Thus it has been essential to explore the nature and shape of the socioeconomic system as it may constitute SI, and how, why, and when portions of the new space industries may arise.

This examination of the terrestrial background has been done in two parts. First, basic macroeconomic projections were made to examine the needs of the human socioeconomic system during the coming decades with respect to basic materials: energy fuels, minerals, and basic agricultural commodities. If the "limits to growth" hypothesis should prove to be correct, then perhaps SI could provide some of the very basic needs of the industrialized societies of the world. Second, a variety of alternative futures were examined to determine how SI might be shaped by events and developments in the rest of the system. The economic profitability, political viability, and social desirability of specific space industrial activities can only be defined in the context of general social, political, economic, and technological factors characterizing an alternative future. These alternative future scenarios also provide some basis for contingency planning and for identifying stepping stones in space technology which are most likely to be useful in any future space programs or activities.

The results of these examinations of the terrestrial background provided some of the basis for considering the market potentials of various possible space industries and much of the foundation for developing specific examples of possible programs of space industrialization during the next few decades. The necessity for continuous planning of intermediate and long-range programs became quite clear from this work.

The following subsections present the results of several examinations and extrapolations used in fixing the terrestrial framework.

A. Resources Assessments

A detailed assessment (case by case) of natural resource availability was made for 18 minerals selected either because of their large volume (such as iron) or because of critical importance to industrial processes or agriculture (such as phosphate). Fossil fuels were also examined in assessing likely sources of energy in the next 3 decades. But to assess supply and demand for such commodities, it was necessary to project population growth and trends in basic economic indicators such as GNP and personal incomes. In addition, the outlook for production of a number of basic agricultural commodities was also examined in the course of these studies.

It does not appear, on the basis of these examinations of energy, minerals, and food production issues over the period 1980 to 2010, that any of these will pose any critical threat to the survival of industrial civilization. The spectre of impending scarcities does not, therefore, provide a credible basis for the political support necessary to mount a major thrust into space at public expense on a crash program schedule. The importance of long-range solutions to the problems of energy supply, however, is clear. The economic value of energy and minerals imported from space may be significant and may provide sufficient motivations for SI; these possibilities should not be dismissed lightly. But their absolute justification, during the period of interest in this study, must be found elsewhere. "Limits to growth" cannot justify SI during the next 3 or more decades.

B. Alternative Scenarios for the Terrestrial Background

Ten events or developments which appear to have a reasonable chance of occurring in the next 2 or 3 decades and which would be likely to have major effects on the shape of the future were used to guide the extrapolation of the terrestrial background. Arranged according to morphological categories, these "triggering" events or developments are as follows:

Extrapolative—Baseline (for comparison-no "triggering events")

Political—Major advances in space by other nations. U.S. commitment to space.

Societal—Major breakthrough in human longevity. U.S. disenchantment with space.

Economic—Entrepreneurial exploitation of space technologies. "Artificial" shortages of critical minerals. Economic collapse due to shortage of capital.

Technological—Breakthrough in a new energy source.

Environmental—Human-generated ecological catastrophies. Rapid cooling of the Northern Hemisphere.

These future scenarios serve several purposes in examining SI. Their formulation and study have provided insight and guidance in expert determination of what future events are most likely to provide opportunity and to precipitate action related to SI. Additionally, they have served to provide the controlling framework used to assess the limits of possible U.S. SI activities and the related economic implications.

The scenarios described here span a wide range of possible futures. While opportunities for the advocacy of a variety of specific SI activities appear in every scenario, many of these opportunities are apt to fall in the private sector rather than in governmental agencies. Because of the strong possibilities of synergisms between various space activities, however, government can no more ignore private sector activities and developments than the private sector can ignore government plans for government of, for example, new launch vehicles. If government's efforts are to have the greatest benefit, those efforts must be based on up-to-date understanding of the opportunties for advocacy presented to the private sector and to the whole public sector by developments in the human system as a whole. This requires continuous examination by government of the changing opportunities and of the changing fabric of the human system. Planning SI cannot be done effectively if it is done only in fits and starts; the volatility of the human system requires reassessment of alternative futures on a continuing basis to identify what space systems and space technologies are most likely to be used by a wide variety of SI opportunities. Just as short-range planning is done on a continuing, day-to-day basis, mid- and long-range planning must be done continuously to prepare for contingencies.

III. INDUSTRIAL OPPORTUNITIES IN SPACE

The establishment of future markets for SI products or services and an SI program for each future scenario requires a knowledge of potential opportunities. These have been established to a level of detail and breadth of application sufficient to allow gross market survey and preliminary program formulation. The result of this has been a compilation of over 200 potential applications for space-related goods and services. These existing lists are by no means exhaustive.

The opportunities and their identifed representative usage have been compiled under four industry activity categories: Information Services, Energy, Products, and People (in space). Each of these categories was further subdivided into subcategories as follows:

Information Services
Communications
Observations
Navigation
Location
Sensor Polling

Energy
Solar Power Satellite
Redirected Insulation
Nuclear Waste Disposal
Nuclear Power/Breeder Satellite
Power Relay

Products
Biologicals
Electronics
Electrical
Structural
Process
Opticals

People
Tourism
Medical
Entertainment/Art
Recreation
Education
Support

IV. POSSIBLE TERRESTRIAL ALTERNATIVES

Thirty-two candidates for space utilization compared to potential Earth based alternatives have been reported.[1] Comparisons were based on examining the initial cost of installation on a first-order basis and a cursory review of qualitative factors such as ease of use, reliability, technology requirements, etc. If costs and capability obtained appeared comparable between the alternatives, they were retained for further study. In certain instances the identified space uses exhibited much lower cost for similar capability or the reverse. These were identified as clearly viable candidates. Where cost and/or capability were clearly superior for the Earth alternative, the candidate was dropped from further consideration.

For 5 of the 32, the terrestrial alternative was deemed clearly superior; 7 appeared more favorably accomplished from space, and 20 depended too much on specific details (too close to call). The following conclusions have been drawn from this assessment. Alternatives do exist, or can be visualized for most space initiatives. "Uniqueness" of the space candidates detailed was not deemed strong enough to warrant special consideration in a competitive environment. Significant technological "lead" for space options was found only in the area of earth resources. And, in the case of communications, implementation may be tipped already toward terrestrial options. In concert with these arguments it is concluded that market softness, in terms of systems requirements, remove the constraint that terrestrial alternative systems must duplicate exactly space products and services.

The implications of the above statements gives rise to the following observations on the viability of terrestrial alternatives:

1. Complexity from detailed assessment of noncost issues substantially reduces the opportunity to develop a "winning" mix of space efforts based on generalized benefits.
2. In lieu of a mandate, space viability must be aggressively advocated/studied against competitors in the mid-1980s.
3. The current involvement of an existing industry will typically indicate which alternative would be favored by it unless forced by competition to change directions. New entries in an industry will select a path based on investment and risk considerations. Most space initiatives considered in this study will appear highly favorable over terrestrial alternatives only after steps toward risk reduction are implemented.

V. POTENTIAL MARKETS AND REVENUES

A. Approach

An extensive list of possible industrial opportunities in SI has been given a cursory market and revenue analysis.[1] These were all taken from subcategories presented in Section III. The resulting data are informative but very preliminary and must be considered in the context of the groundrules and assumptions of their determination.

Different specific methodologies were applied according to which industry was being examined. For example, market analyses for Products were much more speculative than those for Information Services, since much less is known about the specific use and probable cost of a prospective product. A common set of general methodology guidelines was used wherever appropriate and provide the foundation for understanding the philosophy and assumptions which guided these market surveys.

B. Aggregation of Market Potential

One conclusion quickly drawn from the data derived in these analyses is that very large single-opportunity revenues are possible in several areas. It follows that aggregates of these potential (or expanded) industries will represent even great possibilities.

As indicated in each market analysis the flow of revenues initiates and evolves based

Table 1
PROJECTED ANNUAL AND CUMULATIVE REVENUE POTENTIAL FOR SELECTED INFORMATION SERVICES INITIATIVES[a]

	Potential Revenues (in millions of dollars)	
	Annual (Peak)	Cumulative (1985—2010)
Information services		
Pocket telephone	20,000	100,000
Teleconferencing	9,000	90,000
National information services	6,000	40,000
Electronic mail	9,000	90,000
Disaster communications set	30	500
Advanced TV broadcast	2,000	8,000
Vehicle Inspection	300	4,000
Global Search and Rescue	50	300
Nuclear fuel locators	3	40
Ocean Resources	2	50
Transportation services (equipment sales)	70	400
Rail Anti-collision system	40	600
Personal navigation sets (equipment sales)	100	400
Vehicle/package locator	300	5,000
Voting/polling wrist set	40	200
	~$47 billion/year	~$340 billion

[a] 1977 dollars.

on several assumptions including best case/least case for total market potential at saturation. Thus the actual revenues which might be anticipated will depend substantially on the background scenarios previously presented. For purposes of analyzing far and near term implications of SI, the time frame of initiation and rate of growth of each industry was adjusted during the analysis.

Although the more exact figures will depend on such specifics as future scenarios and programmatics, it is worthwhile to summarize revenue projections. This will allow interpretation by the reader of the possible significance of SI in the near future. Summary data are presented in Tables 1 through 4 for revenues (best case). The timing shown is considered to correspond roughly to the three decades 1980 to 2010 and the baseline scenario. All revenues calculated are additive to current SI revenues which total about $1 billion per year in 1978.

The relative value of each market area (Information Services, Energy, Products, and People) as a function of time is presented in Figure 1. A more aggressive scenario (implying more aggressive SI programs) basically accelerates the revenue flow and adds more minor initiatives. A scenario without Solar Power Satellite(s) (SPS) eliminates that portion of the summary and inhibits other activities in the Information Services and Products industries.

Inherent to this prediction is a period of capability and technology development in the 1980s leading to expanded exploitation and utilization in the 1990s and intensive growth beyond the year 2000. The resulting interpretation from this is that revenues will approximately double from 1980 to 1990. A very rapid growth in revenue then ensues as technology and hardware development effort in the 1980s come on line in the 1990 to 1995 time period. Although new technologies (particularly in power, structures, and transportation) are emerging, the activities settle basically into an implementation and expansion phase with doubling time for revenues becoming approximately 5 years until the end of the time period of interest.

Table 2

PROJECTED ANNUAL AND CUMULATIVE REVENUE POTENTIAL FOR SELECTED ENERGY INITIATIVES[a]

| | Potential revenues (in millions of dollars) | |
	Annual (peak)	Cumulative (1985—2010)
Energy		
Solar Power Satellite (first SAT in 1996)		
49 5GW at 27 mils/kWh	50,000	300,000
60 10GW at 11.5 mils/kWh → 7.1 MILS/kWh	30,000	200,000
60 10GW at 27 mils/kWh	100,000	600,000
Urban night illuminator	200	2,000
Nuclear waste disposal	1,000	3,000
	~$30—$100 billion	~$200—$600 billion

[a] 1977 dollars.

Table 3

PROJECTED ANNUAL AND CUMULATIVE REVENUE POTENTIAL FOR SELECTED PRODUCTS[a]

| | Potential revenues (in millions of dollars) | |
	Annual (peak)	Cumulative (1985—2010)
Products		
Drugs and pharmaceuticals	600	7,000
Electronics		
Semiconductors	2,000	20,000
Electrical		
Magnets	300	4,000
Superconductor (generating only)	2,000	20,000
Optical		
Fiber optics	80	800
Special metals		
Perishable cutting tools	800	8,000
Bearings and bushings	200	2,000
Jewelry	100	2,000
	~$6 billion/year	~$64 billion

[a] 1977 dollars.

There appear to be no "natural laws" or technological barriers which would limit revenue growth to the level indicated. Strong response to a foreign challenge or a heavy entrepreneurial initiative in the near future (early 1980s) could result in more rapid growth. Potential revenues well over $100 billion (1977) per year appear feasible with technology and development effort acceleration and aggressive marketing.

VI. POTENTIAL SPACE INDUSTRY PROGRAMS

The results of the work described in previous sections (Terrestrial Framework, Opportunities, and Markets) provided the basic information necessary to map out

Table 4

**PROJECTED ANNUAL AND CUMULATIVE REVENUE POTENTIAL FOR
SELECTED PEOPLE INITIATIVES[a]**

	Potential revenues (in millions of dollars)	
	Annual (peak)	**Cumulative (1985—2010)**
People		
Space tourism	50	900
Space hotel	50	600
	~$100 million/year	~$1.5 billion

[a] 1977 dollars.

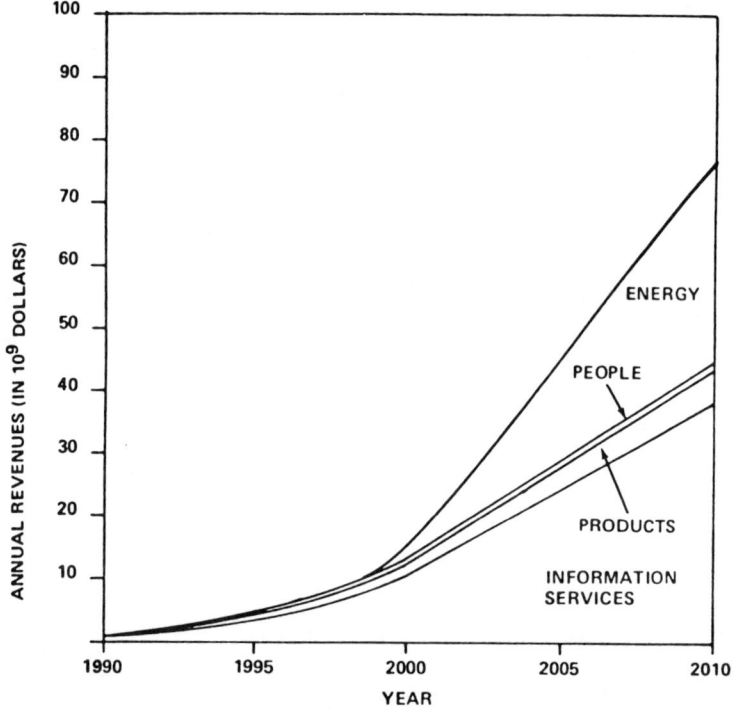

FIGURE 1.   Projected revenues for space industry activities assuming the baseline scenario for terrestrial framework.

probable courses of future private and government programs. SI by definition consists of a multitude of development or operational programs occurring simultaneously or in series.

The flow and interrelationship of various information from which programs have been derived and analyzed is shown on Figure 2 which also summarizes the grouping of 11 scenarios into 6 possible programs.[1]

The hypothetical programs shown reflect considered collective judgement of what kinds of space activities are likely to be economically and politically viable in the assumed contexts of the background scenarios.[1] Like the scenarios themselves, these programs should not be interpreted as attempts to forecast the future but rather as

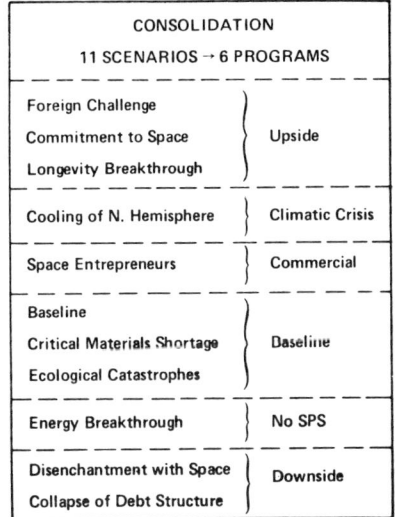
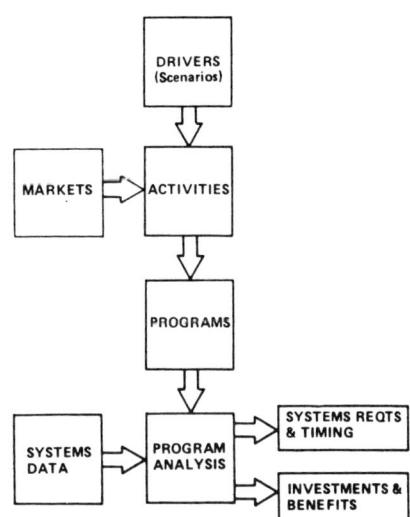

FIGURE 2. Summary of space industry programs derived and the steps to their derivation and analysis. All values shown are additive to the present (1977) revenues of approximately $1 billion/year. All figures are in 1977 dollars.

explorations of plausible alternatives which can be used to provide reasonable guidelines for long-range planning. This kind of exploration can serve to identify stepping stones which are common to a multitude of alternative futures and to identify systems which are likely to be much more dependent on external contingencies, requiring more careful attention to the course of events.

To scope future costs, impacts, and requirements, three programs were selected for detailed analysis. These were the Baseline, No SPS, and Upside Programs. The procedure was to take the general programs and adjust specific timing and scale according to guidelines from the scenarios and market data. In some cases several initiatives were lumped into a single system for purposes of simplifying costing and scheduling.

Introduction and growth of the various Information Services leads to requirements for very large platforms in the 1990s and beyond, corresponding to the market projections presented previously. This turned out to be the largest single set of industry initiatives in terms of power and structures other than SPS and light reflectors.

The largest single need for mass in Earth orbit was determined to be the SPS. The light reflectors for illuminating large areas of the Earth were the next largest mass requirement where a large number was needed. The third largest mass-need identified was for materials-processing in space, based on the assumption that future costs to provide the raw materials will decrease by a factor of 10 to 20 when compared to the shuttle. All three of these mass requirements can potentially economically justify the conversion from Earth based to nonterrestrial (lunar or asteroidal) materials.[1,4,6,7] This aspect of SI will be discussed in more detail in later chapters.

VII. WORLDWIDE ACTIVITIES IN SI AND THEIR IMPLICATIONS

From military and scientific beginnings some 20 years ago (October 1957 for Sputnik, January 1958 for Explorer) there has evolved a broad and complex industrial base in space. The activities range from basic research in the space processing of materials to the fully operational information transfer systems. The worldwide gross annual revenue now exceeds $1 billion in sales of services alone. Current published projections indicate that revenues from services by 2010 may reach $10 to 20 billion given only minor extrapolations of present technology. With technology advancements in power, structures, transportation, materials processing, frequency utilization, and data handling, the potential can be several times that revenue amount. Of the four general categories of SI (Information Services, Products, Energy, People in space) the area nearest maturity is Information Services. New technologies will be necessary to open up new markets in these services also, however.

The worldwide interest in SI is reflected by the number of countries and agencies that are actively participating at present. This is characterized in Figure 3 by summarizing capabilities both previously demonstrated and currently being developed (as of 1977) that relate specifically to SI.

Why is there such extensive involvement in organizing for and implementing SI in the world? In the simplest terms, it appears that needs and markets exist forming the basis for large scale international involvements. This has prompted a wide-spread interest and desire for independent capabilities to utilize space and an awareness of the potential benefits from gaining and maintaining a competitive position. In the free world the U.S. will be challenged through the 1980s in all technologies including those that are peculiar to manned space flight. The recent capability demonstrations of the U.S.S.R. aboard the Salyut space station and the strong reports of their current development of a reusable shuttle, leave no doubt that major technical achievements can be anticipated by communist bloc countries in SI throughout the 1980s.

VIII. DOMESTIC IMPACTS AND ISSUES

SI will have impact and meaning in many areas of human endeavor in the U.S. as well as the world. These impacts may be more subtle than profound, given the level of economic and technical development already enjoyed by the U.S. Nonetheless, these impacts will be important and, indeed, can provide the stimuli necessary to assure U.S. economic growth through the last decade of this century and into the next. The realization of these impacts in a positive fashion will depend completely on establishment of government roles supportive to greatly enhanced private sector involvement in space.

A. Employment and Economy Impacts

The economic impacts examined in Reference 1 were based on an alternate approach to simple trend extrapolation. From the scenario, market and program work reported, it was surmised that significant opportunities existed in the near future in the public and private sector. If advantage is taken of these opportunities, a much greater benefit to the U.S. and world can be realized than would be predicted by simple trend forecasting. Thus the analyses were structured to show what could happen, and hopefully help precipitate the required actions. As mentioned in the discussion on Future Scenarios, an implicit assumption in all analyses was that opportunities would be capitalized on, the extent of which varied between scenarios.

1. Assumptions

Certain assumptions concerning the nature of the industries involved in stimulating and handling the revenue flows which provided the bases for economic analysis were necessary. Composite indices were developed and applied across the board without specific variation between industry types, products vs. services, etc. Given the

NATION	GENERAL							UNMANNED		MANNED			UNMANNED			MANNED			
	DATA HANDLING	GROUND STATION(S)	LAUNCH FACILITY (S)	SUBORBITAL LAUNCH	LEO LAUNCH	LEO RETURN	GSO LAUNCH	ORBITAL RENDEZVOUS	LEO LAUNCH	LEO OCCUPANCY	ORBITAL RENDEZVOUS	ORB PROPELLANT TRANS.	EARTH OBSERVATION	COMMUNICATIONS	NAVIGATION	TEST & EXPERIMEN.	MAINT. & REPAIR	REMOTE CONTROL	MATERIALS PROCESSING
UNITED STATES	X	X	X	X	X	X	X	X	X	X	X		X	X	X	X	X	X	X
USSR	X	X	X	X	X	X	X	⊗	⊗	⊗	⊗	⊗	X	X	X	⊗	⊗	X	X
CHINA (PR)	X	X	X	X	X	X							X	X					
FRANCE	X	X	X	X	X	?							X	X					X
INDIA	X	X	X	X	(X)								X	X					
JAPAN	X	X	X	X	X									X					
ESA	X	X	X	X	(X)		(X)			(X)			X	X		(X)			X
OTHER	X	X	X	X	X								X	X					X
TOTAL NUMBER OF NATIONS	111	39	24	15	9	3+	4	2	2	3	2	1	7	13	2	3	2	3	5

(X) Indicates to be demonstrated by 1981   ⊗ Indicates currently unique capability
X Indicates capability has been demonstrated.

FIGURE 3. Worldwide SI capability summary. (X) indicates to be demonstrated by 1981; ⊗ indicates currently unique capability; X indicates capability has been demonstrated.

composite nature of the expenditures and revenues on a whole program basis and our uncertainties, even detailed indices would yield aggregate approximations. Thus a more detailed industry by industry assessment is not warranted until more specific analyses are performed on the projected revenues and costs.

The specific indices used were

Before tax profit = 0.20
    Gross revenue
Tax bracket = 0.50
Labor intensity (operations and maintenance) = 0.40
Composite mean salary = $17,000 annual

These figures are typical of various service industries today subject to government regulation. Although profit margin for COMSAT was somewhat higher in the past, the trend has been to force it downward. As discussed in Reference 1, after tax profits have varied from 6 to 15% typically. Current corporate taxes are 49%. For simplicity, a straight 50% was used.

Calculations were performed for two points in time. First, 1985 was selected as a representative near-term year where almost all new revenue in SI would be investments. This was assumed to be almost purely public funding (government sponsored research and development programs). The latter year chosen was 2010, the end of the time period being examined (1980 to 2010). The revenue in 2010 is projected to result almost entirely (>95%) from sales of Services, Products, and Energy (People in Space revenues are insignificant compared to the others.)

2. Results

Jobs and taxes generated were estimated for three programs: the Energy Breakthrough (No SPS), the Baseline (with SPS), and the Upside (all initiatives).

Although the Upside is considered the least likely of the three (requiring heavy investment in the early 1980s), it was desirable to assess the potential impact of such a strong, aggressive set of initiatives. The results were as follows:

NEW JOBS[a]

	No SPS	Baseline	Upside
1985	15,000	100,000	120,000
2010	1,000,000	1,900,000	3,800,000

TAXES GENERATED[a]

	No SPS	Baseline	Upside
1985	$ 100M	$ 800M	$ 1,000M
2010	$10,000M	$20,000M	$40,000M

[a] Direct only. U.S. markets only.

The estimate of jobs for 1985 is probably low by a factor of two since most funding would be to the aerospace industries. The Aerospace Industries Association (AIA) has estimated that about 30 direct jobs are created for each $1 million of appropriation. Direct plus indirect jobs are estimated to total about 100 jobs per $1 million. Thus the job projection for 1985 is conservative since the computation was the same as for 2010. The true impact on new jobs is some two to four times the numbers shown here depending on specific assumptions and economic theory applied.

In the aggregate, the best guess is that 75% or more of the postulated SI initiatives revenues will be job creating in the 1990s and beyond. Thus for a workforce of 100 million in 2010 some 3 to 12% could be employed in new jobs created by SI.

The tax revenue calculations take into account corporate taxes based on previous assumptions plus personal income taxes of direct employees. A national composite rate of 0.26 for federal and state income tax was applied to personal income.

B. Sample Industry Comments and Recommendations

Several reviews of investment possibilities for private industry and investors have been conducted in the last 3 years. The following comments, observations, and conclusions have been drawn from those reviews.

At present a corporate leader/manager cannot justify any investment in products (at the basic research stage and having much too high transport costs) or energy (overwhelming techno-economic risk). The more expensive communications systems (such as the Orbital Antenna Farm of Morgan and Edelson, COMSAT) are being looked at rather seriously, although the total capital and payback times on larger systems look doubtful.[1] One simple message is certain. As techno-economic risks come down, U.S. industry will steadily increase its allocation of resources to SI if a reasonable payback can be obtained. "Reasonable payback" will vary broadly based on initial investment, near term vs. long term risk, guarantees, etc.

1. Industry Views on Roles and Responsibilities

The specifics of appropriate roles and responsibilities which could be adopted by industry and government vary broadly according to the industry and the individual. It is possible, however, at the rather gross segregation level presented in Figure 4 to assemble a set of consensus opinions. As might be anticipated the communications industry is sufficiently mature that the Product Development and Pilot Operations

INDUSTRY IDEAS ON ROLES & RESPONSIBILITEIS

C — COMMUNICATIONS
E — ENERGY
P — PRODUCTS

	GOVERNMENT ACTIVITY	INDUSTRY ACTIVITY	JOINT VENTURE	GOVERNMENT REGULATION
BASIC RESEARCH	C E P			
APPLIED RESEARCH	E P		C	
PRODUCT DEVELOPMENT	E	C*	C* P	
PILOT OPERATIONS	E	C*	C* P	
PRODUCTION OPERATIONS		C E P		C E
TRANSPORT DEVELOPMENT	C E P			
TRANSPORT OPERATIONS		C E P		C E P

SAI-4242

*DEPENDS ON SPECIFICS

FIGURE 4.   Industry views on roles and responsibilities.

areas require consideration of specific proposals to obtain a particular opinion. The large communications platform concept was one initiative that generally fell in the joint-venture category for example. A new version of an existing satellite system was considered to be an appropriate industry activity. Particular attention is drawn to the concensus or CEP blocks.

The information presented here is considered as a stage setting providing general guidelines for development of specific arrangements on a case by case basis. Early general agreement to these guidelines by government would encourage enhanced industry involvement in SI.

Specific concerns and recommendations from industry have been compiled into a rather extensive list. These are presented in Volume 3 of Reference 1. The following policy on NASA/industry relations (dated June 25, 1979) regarding SI recognizes many of these concerns and recommendations.

2. NASA Guidelines Regarding Early Usage of Space for Industrial Purposes
NASA, by virtue of the National Aeronautics and Space Act of 1958, is directed to conduct its activities so as to contribute to the preservation of the role of the U.S. as a leader in aeronautical and space science and technology and their applications.

Since substantial portions of the U.S. technological base and motivation reside in the U.S. private sector, NASA will enter into transactions and take necessary and proper actions to achieve the objective of national technological superiority through joint action with U.S. domestic concerns. These transactions and actions will be undertaken in the context of stated NASA program objectives and after a determination by the administrator. They may include, but are not limited to: (1) engaging in joint arrangements with U.S. domestic concerns in research programs directed to the development of enhancement of U.S. commercial leadership utilizing the space environment; (2) conducting research programs having as an end objective the enhancement of U.S. capability by developing space-related high-risk or long-lead-time technology; and (3) entering into transactions with U.S. concerns designed to encourage the commercial availability of products of NASA space flight systems.

NASA incentives for these purposes may include in addition to making available the results of NASA research: (1) proving flight time on the space transportation system on appropriate terms and conditions as determined by the Administrator; (2) providing technical advice, consultation, data, equipment, and facilities to participating organizations; and (3) entering into joint research and demonstration programs where each party funds its own participation.

In making the necessary determination to proceed under this policy, the administrator will consider the need for NASA-funded support to commercial endeavors and the relative benefits to be obtained from such endeavors.

As major areas for NASA enhancement of total U.S. capability, including the private sector, may become apparent from time to time, the factors to be considered by NASA prior to providing incentives may include, but not be limited to, some or all of the following considerations:

1. The public or social need for the expected technology development
2. The contribution to be made to the maintenance of U.S. technology superiority
3. Possible benefits accruing to the public or the U.S. government from sharing in results
4. The enhanced economic exploitation of NASA capabilities such as the space transportation system
5. The desirability of private sector involvement in NASA programs
6. The merit of the research, development, or application proposed
7. The degree of risk and financial participation by the commercial concern
8. The amount of proprietary data or background information to be furnished by the concern.
9. The rights in data to be granted the concern in consideration of its contribution
10. The ability of the concern to project a potential market
11. The willingness and ability of the concern to market and sell any resulting new or enhanced products on a reasonable basis
12. The impact of NASA sponsorship on a given industry
13. Provision for a form of exclusivity in special cases when needed to promote innovation
14. Recoupment of the NASA contribution under appropriate circumstances
15. Support of the socio-economic objectives of the government

IX. LEGAL CONSIDERATIONS

The legal considerations which must be recognized in any broad examination of SI are, at first glimpse, staggering in their number and complexity. Entire symposia and extensive sessions during various astronautical conferences have been devoted to deliberations on interpretation of existing and proposed agreements, treaties, statutes, etc. Expert opinion has recently been expressed on two questions: Are there any international laws or agreements which would preclude or severely limit evolution of any of the SI initiatives discussed in this study? Are there domestic laws (in the U.S.) with similar potential for limitation?

At present, SI is not suffering substantially from either international or domestic legal constraints. The large-scale initiatives discussed in this publication for implementation in the 1980 to 2010 period could all be exercised today within the legal structure. However, all indications are that a series of initiatives to limit the U.S. and its industry are in existence or are being originated. A net effect of these in the light of no established national policy in this arena will be to increase economic risk and foster impediment to industry involvement in SI. Without steps to assure industry in these matters, SI may falter regardless of economic and technological enticements.

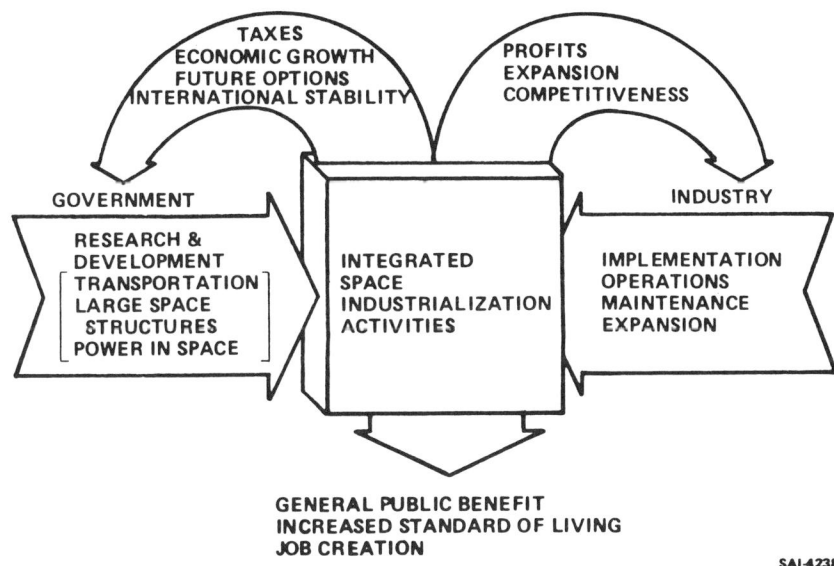

FIGURE 5.   The how and why of SI.

X. INSTITUTION IMPLICATIONS

A host of institutions ranging from religious to technological will be affected by SI. Their influences on the growth of SI and the specific opportunities which will be capitalized on vary greatly, of course. Based on comments by industry leaders, investors, lawyers, and national politicians there are basic considerations which form the backdrop for detailed analysis of key institutional implications.

These "Five Significant Considerations" are

1. SI must become an integral part of national space policy planning.
2. Industrializing organizations and legal structures must evolve and be encouraged.
3. Mechanisms for advantageous transfer of responsibility will be necessary.
4. The applicability of SI technologies to many problems, needs, and markets will go unnoticed without focused dialog.
5. The knotty issues of today in technology export will be further driven by the international/multinational nature of SI.

The government/industry/academic institutional arrangements necessary to accomplish items 1 through 3 and optimize benefit from 4 and 5 must be designed in a fashion responsive to the needs of individual SI initiatives.

XI. OBSERVATIONS

Figure 5 summarizes the how and why of SI. The roles and related activities of government and industry feed the integrated SI activities that represents the summation of all future private and public SI programs. The motivators for such input and sponsorship are shown as a feedback loop. Three encompassing benefits to the public at large are shown as the integrated "value generation" of space industrialization. Net value generation is possible because a new and virtually inexhaustible resource, loosely called "space", is being utilized.

The corollary consideration of what and when are primarily addressed in subsequent chapters. At the highest aggregation level (general industrial category) the what and when can be displayed as in Figure 1. The rather sterile lines on this figure are wholly inadequate to express the latitude that each has according to the scenario assumed and the realities of capitalization and investment. Also the natural or "organic" growth potential of existing space industry (such as Intelsat and COMSAT) are not reflected in Figure 1. Thus the potential exists for the curves of each industry to swing toward greater or less revenue or benefit. We can say with some confidence that the direction and magnitude of that swing will depend completely on the investments made in the 1980s. The technological, legal, institutional, international, and regulatory hurdles must all overcome in concert. The most important, however, is technological. The incentives are sufficiently strong that, given technical capability, the other hurdles will be overcome. The future of the U.S. and our system of free enterprise and democracy demand it. Therein lies the challenge to the U.S., the free worlds technological leader in space.

The most important spur to U.S. industry and U.S. world technical leadership will result from near term development of the following key technologies.

1. Large information systems
   - Structures—Large antenna of 10 m or 200 m diameter
   - Power—20 kW to 10,000 kW
   - Data processing—100 to 1000 times present rate
   - Transportation to high orbit—routine for maintenance, repair
2. Materials space processing
   - Laboratory demonstration—goal oriented spar, spacelab
   - Prototype production—10 to 1000 lb/day on some products
   - Orbital support systems—power, structure, stability
   - Low cost transportation to low orbit—<$100/lb to really open market
3. Large energy systems (use in space, broadcast to Earth)
   - Structures—0.5 km to 15 km
   - Power handling—25 kW to 10 gw
   - Low cost transportation to high orbit—minimum feasible cost

The projected benefits depend upon commercial operations that can only begin after the key technologies are available. The potential benefits are significant covering a spectrum of national concerns from jobs and balance of trade through standard of living and national pride.

Studies have shown that commercially viable industries in Information Services, Energy, and Products can be realized given the tools discussed in Figure 6 and illustrated in Figure 7 for private enterprise to work with. Analyses have also shown that public investment in these capabilities in he 1980s will be paid back manyfold in the 1990s and beyond. All indcations are that SI is the stimulus required to swing the U.S. and the world upward toward the next plateau of human achievement. This will be achievement *in toto,* not just in space or on the Earth but through the sphere of human endeavor.

SI is, as Ehricke termed it, the overarching concept capable of encompassing and coordinating future applications of space in the most beneficial manner. The individual initiatives in the industrial activities identified in this study should not be pursued as autonomous projects unto themselves. Study has shown that the concepts of SI are viable; and that industry/government cooperative planning and implementation are desirable and feasible.

Space Industrialization exists and will grow. How rapidly and to what saturation level are the key questions.

ACTIVITY	INFORMATION	ENERGY	MATERIALS	PEOPLE
MAJOR SPACE ADVANTAGE	• VIEW • ACCESS	• SOLAR FLUX	• LOW 'G' HIGH VAC • HIGH VAC	• UNIQUENESS
MAJOR TECHNICAL HURDLES	• SIZE 10–100 METER ANTENNA • POWER 21 KW – 10,000 KW • DATA PROC • TRANSPORT COST (OPERATIONS)	• SIZE/MASS OF SYSTEM ~ 10^4 MW ~ 10^5 TONS ~ $\$10^{10}$ • TRANSPORT COST < \$20/LB LEO • ENVIRONMENT ISSUES	• PROOF OF THEORY • PRODUCTION DEVELOPMENT HUNDREDS OF POUNDS PER DAY • POWER 10 KW – 10,000 KW CONTINUOUS • TRANSPORT COST < \$100/LB LEO	• TRANSPORT COST \$25/LB OR LESS • HABITATION
TIMING FOR SIGNIFICANT REVENUES	• PRESENT > \$1000 M/YR • 1985 + RAPID EXPANSION	• 1996 +	• 1987 +	• 1990 +

FIGURE 6. A summary of qualitative and quantitative observations drawn from the programs analysis.

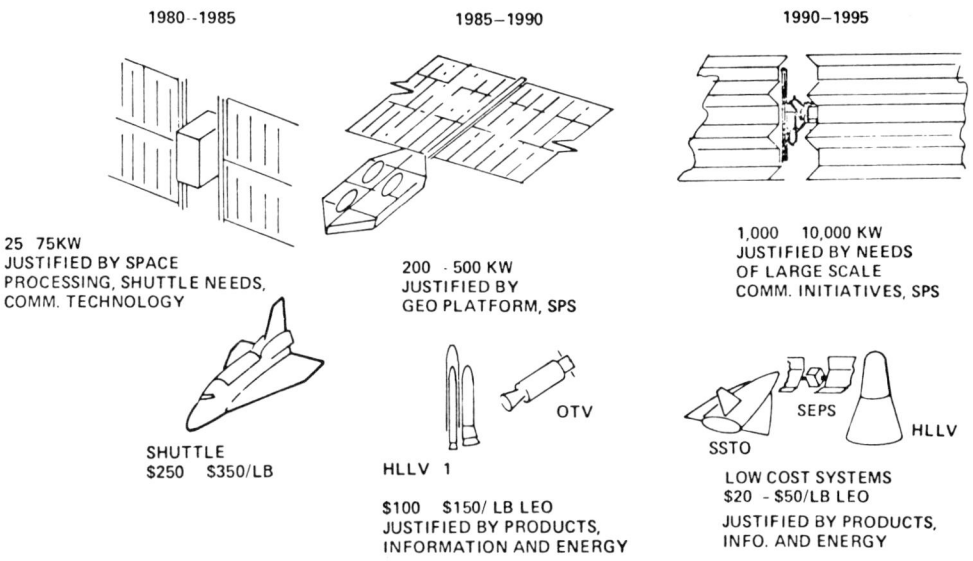

FIGURE 7. SI structures, power, and transport requirements.

REFERENCES

1, Space Industrialization Study—Final Report, NASA Contract NAS8-32197, Science Applications, Inc., Huntsville, Ala., April 1978.
2. Space Industrialization—Final Report, NASA Contract NAS8-32198, Rockwell International/Space Division, Downey, Calif., April 1978.
3. Davis, H. P., Power from Space—A New Opportunity, in *Proc. 71st Annu. Mtg. Am. Inst. Chem. Engineers,* November 1978.

4. Space-Based Manufacturing from Non-terrestrial Materials, in *Progress in Astronautics and Aeronautics,* Vol. 57, American Institute of Aeronautics and Astronautics, 1977.

5. Brown, R., Discussions of materials processing in space with corporate representatives, in Briefing hosted by American Institute of Aeronautics and Astronautics, New York, October, 1978.

6. Lunar Resources Utilization for Space Construction—Final Report, NASA Contract NAS9-15560, General Dynamics/Convair Division, San Diego, Tex., 1979).

7. Driggers, G. W., Is lunar material use practical in a non-SPS scenario?, *Proc. 4th Princeton/AIAA Conf. on Spce Manufacturing Facilities,* Grey, J. and Krop, C., Eds., ATAA Press, October 1979.

Chapter 2

NEAR TERM PRODUCTS AND SERVICES

Gerald W. Driggers

TABLE OF CONTENTS

I. INTRODUCTION

The products and services to be examined in this chapter are those considered technologically achievable on a full-scale basis in 15 years or less, although this definition of "near term" is somewhat arbitrary. This time frame was selected because projections of what can be achieved are typically overoptimistic in the very near term (5 to 10 years) and pessimistic in the far term (25 to 50 years). This is not to say that all things discussed here are predicted to occur; only that they appear to be technologically achievable.

Two general market locales already exist for SI products and services: one on Earth and one in space. The space-based market for materials, finished parts, power, and maintenance will grow as the Earth-based demand grows. Thus a service demand on Earth will create a product and service demand in space. Also, a product demand on Earth will create a product and service demand in space. As indicated in Chapter 1 on Overview, these demands can be of substantial size.

As discussed in a later section of this chapter and in other chapters, when the space demand is large enough a new industry base using raw materials of nonterrestrial origin becomes practical economically.[1,2] That intriguing aspect of near term products and services will also be examined. Of principal near term interest, however, are the possibilities for Earth and shuttle-based increases in service industry scale and initiation of low Earth orbit commercial materials processing.

II. INFORMATION SERVICES

The current worldwide activities in satellite communications represent an essential element of the human socioeconomic system. Many countries would literally be cut off from the world if the current systems ceased to function. Also, the level of international communications would fall dramatically and associated costs would rise. Economic impact would be substantial in some sectors, since on the order of a billion dollars a year is expended on national and international communications.

This is, however, but a small indication of what is to come. As discussed in Chapter 1, growths of 10 to 100 times current revenues can probably be expected in 20 to 30 years.[3] This implies significant growth in communications satellite technologies during the next 15 years to allow geosynchronous orbit to serve the needs of the world beyond 1995. The precise nature of this technological growth is as yet uncertain. The large multipurpose platform seems rather a certainty, especially if the benefits of manned or remote manipulator maintenance and repair prove attractive. The following concept of such a large, multipurpose platform is presented as an example of what could evolve as an early step leading to larger, more powerful systems.

To provide some basis for the design of this platform, Rockwell International (RI) in Reference 4 drew upon Reference 3 for guidance. The level of business anticipated and projected near term technology were used as a guide for selection of which services to offer. The more extensive list of total services considered is presented in Chapter 1. Detailed discussions of most of the services listed can be found in the work of Bekey presented in Reference 3.

Five initiatives in Information Services represent over 90% of the potential domestic revenue from space in Information Services (see Chapter 1). These are (1) Portable Telephone, (2) Teleconferencing, (3) National Information Service, (4) Direct Broadcast TV, and (5) Electronic Mail. The large potential market for these services was taken as a natural indicator of private investment potential.

A representative multiuse Geosynchronous Platform to provide market entry into the five services is shown in Figures 1 and 2.[4] This design provided the basis for

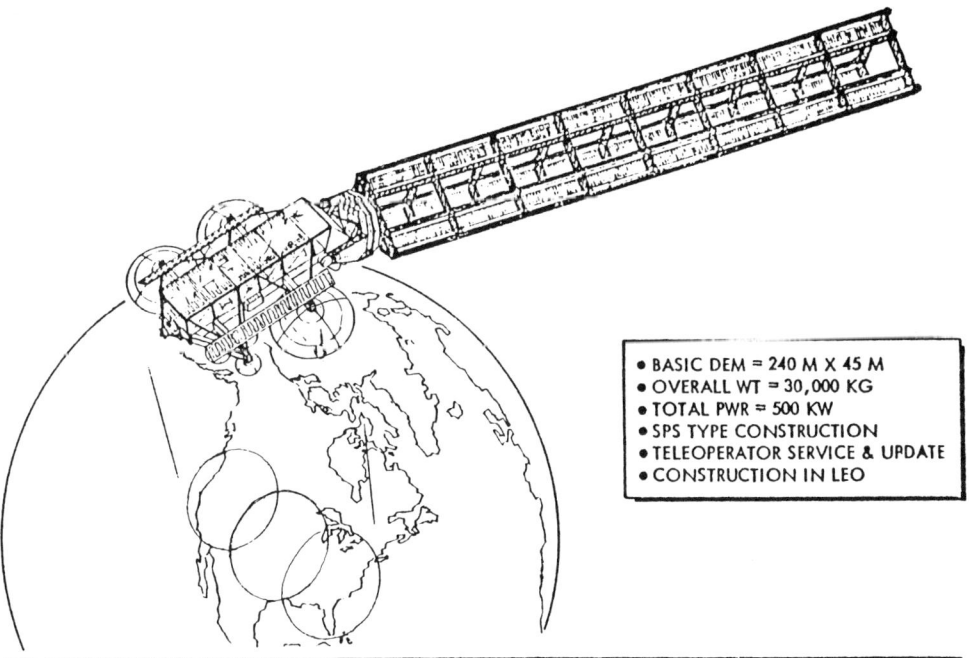

- BASIC DEM ≈ 240 M X 45 M
- OVERALL WT ≈ 30,000 KG
- TOTAL PWR ≈ 500 KW
- SPS TYPE CONSTRUCTION
- TELEOPERATOR SERVICE & UPDATE
- CONSTRUCTION IN LEO

FIGURE 1.   500 kW Geosynchronous Platform.

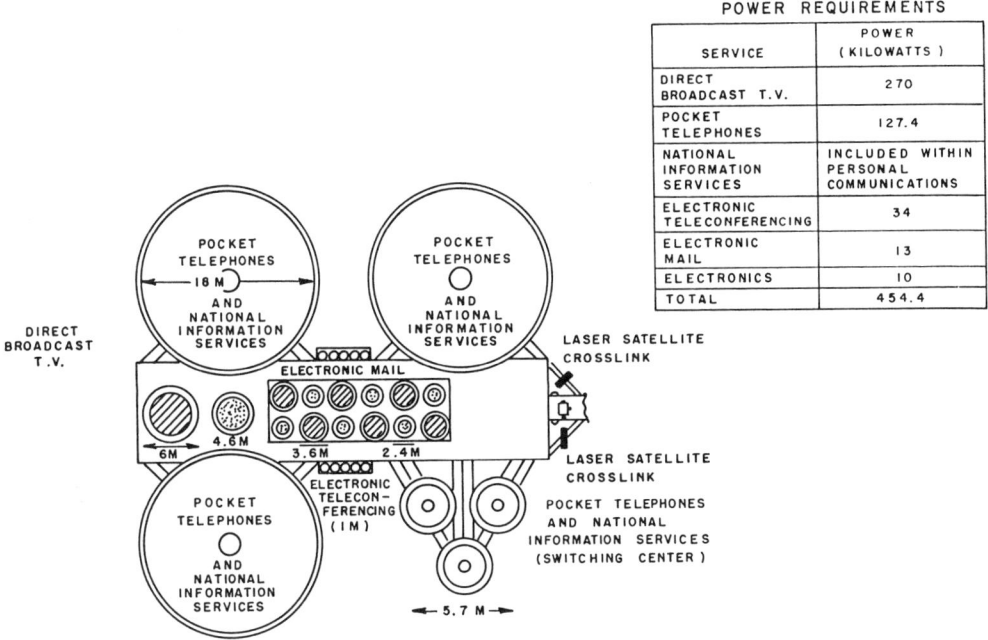

POWER REQUIREMENTS

SERVICE	POWER (KILOWATTS)
DIRECT BROADCAST T.V.	270
POCKET TELEPHONES	127.4
NATIONAL INFORMATION SERVICES	INCLUDED WITHIN PERSONAL COMMUNICATIONS
ELECTRONIC TELECONFERENCING	34
ELECTRONIC MAIL	13
ELECTRONICS	10
TOTAL	454.4

FIGURE 2.   Antenna locations for the Geosynchronous Platform.

development and initial operational cost assessment. As the markets develop into the 1990s and beyond, larger versions of the same type system will be required.

The specific functions provided by this early satellite are as follows:

Direct broadcast television—Five simultaneous color video channels 16 hr/day to be received on modified, conventional TV receivers. The entire CONUS area is to be covered (excluding Alaska, Hawaii, and Puerto Rico).

Pocket telephones—Multiple voice channels originating from remote, wireless extensions; to be connected to conventional fixed terminals or other remote terminals via satellite. Saturation capacity to be at least 45,000 simultaneous transactions.

National information services—Direct access via satellite from home or business intelligent terminals to computer-supported data banks, such as the Library of Congress.

Teleconferencing—Two-way (or multiple) video links between as many as 30 ground sites simultaneously (150 conferences). User locations would have studio-type facilities, including multiple cameras and monitors, switch gear, and communications.

Electronic mail—Facsimile transmission of personal and business correspondence. Terminals would be located in regional postal centers. The regional centers would be interconnected via satellite. Each regional postal center would contain equipment to convert hard copy to electronic facsimile, and vice versa. The ultimate goal is to delivery 40 million pages (8½ × 11) from source to destination overnight.

The level of business projected for penetration by these space based services was presented in Chapter 1 as revenue vs. time. The exact levels reached by 1995 will depend on how aggressive the satellite services proponents and business interests are in the 1980s. Only a few (2 to 5) of the platforms illustrated here would be required to meet the market demands. It is essential, however, that the technology to transport, build, power, control, and maintain such large service systems be developed in the 1980s if commercialization and growth in the 1990s is to occur.

The import of the last statement is illustrated by Figure 3 where a longer term projection of satellite requirements in presented. The two initiatives examined in detail to generate these data were the portable telephone and teleconferencing. These growth projections are based on initial market penetrations in the early 1990s using the baseline scenario described in the previous chapter as a guide to project total market and penetration rate. The toal information services will require substantially more mass and power for satellites and all related technologies will of course advance beyond 1995. The essential technology base will be built from 1980 to 1990, however, if the initial benefits of SI are to be realized in the 1990s.

III. PRODUCTS

The potential to use the unique environment of space (microgravity or microgravity in conjunction with high pumping rate vacuum) for generating unique knowledge of products has been of interest for over 12 years. Materials processing in space (MPS) research has become a funded activity in all countries currently sponsoring space activities including the U.S.S.R., Japan, the European nations and China as well as the U.S. Concerted efforts are now underway in the U.S. to develop the fundamental technologies and business relationships which will lead to commercialization of MPS products in the late 1980s to early 1990s using the space shuttle. This section will summarize the history and status of MPS and provide information on commercialization activities.

A. History

The ground-based research leading to experiments in space began in the late 1960s primarily at the NASA Marshall Space Flight Center in Huntsville, Ala. Drop towers (1 to 3 sec of μg) and aircraft (10 to 30 sec of μg) were used for initial experiments. The first space-based orbital experiments came in 1971 with Apollo 14. A total of five experiments flew aboard Apollos 14, 16, and 17.

A much more sophisticated set of experiments flew aboard Skylab in 1973 and 1974. Some 23 experiments were programmed and a large part of the current technical base in MPS came from their results. The last manned U.S. MPS effort prior to space shuttle operations came in 1975 during the joint U.S./U.S.S.R. Apollo/Soyuz. Some 13 experiments were conducted jointly by the U.S. Astronauts and U.S.S.R. Cosmonauts.

Since Apollo/Soyuz, the U.S. has been limited to the unmanned Space Processing Application Rocket (SPAR) Project using the Black Brant sounding rocket. Low gravity operation of 5 to 7 min are provided.

The following listing summarizes the work done during these various programs. In total, some 60 experiments involving about 100 different materials scientists have been performed.

1971—Apollo
5 experiments demonstrated
Heat flow and convection
Electrophoretic separation
Composite casting

1973—Skylab
23 experiments explored

Bio-separation
Composites
Eutectics
Immiscibles
Electron beam welding
Exothermic bracing
Containerless melting

1975—Apollo-Soyuz
13 experiments explored
Electrophoresis
Surface tension
Magnetic materials
Crystal growth
Metal composites
Monotectics
Eutectics

1976 to 1977—Spar
28 experiments in these
experiment disciplines
Metallurgy
Electronic materials
Fluid mechanics
Bio preparations
Blass/ceramics

The space allowed here will not permit descriptions of experiments and results to date. These things are, indeed, not appropriate to this volume and deserve an extensive treatment of their own. Much has been learned concerning apparatus, procedures, and materials behavior. Experiment time and volume, mass and power limitations have restricted much of the work and hampered the results. The current activities reported in the next section are being designed to use the space shuttle to advantage in overcoming these.

An extensive bibliography (containing several hundred entries) on MPS is available from the MPS project at Marshall Space Flight Center, Huntsville, Ala. Equipments, experiments, results, and commercialization studies are all addressed.

B. Current Activities

The presently sponsored work in MPS is oriented toward a series of experiments to be conducted using the space shuttle in the 1981 to 1983 time frame. Equipment development, particularly for use in Spacelab (a Space Shuttle payload laboratory being developed by the European Space Agency) and ground-based experiments are currently in progress. The following list summarizes current areas of research being conducted in government laboratories, universities, research institutions, and in industry.

1. Crystal growth and solidification
   A. Solid solution in detectors (HgCdTe, PbSnTe)
   B. Vapor growth (HgI$_2$, alloy type

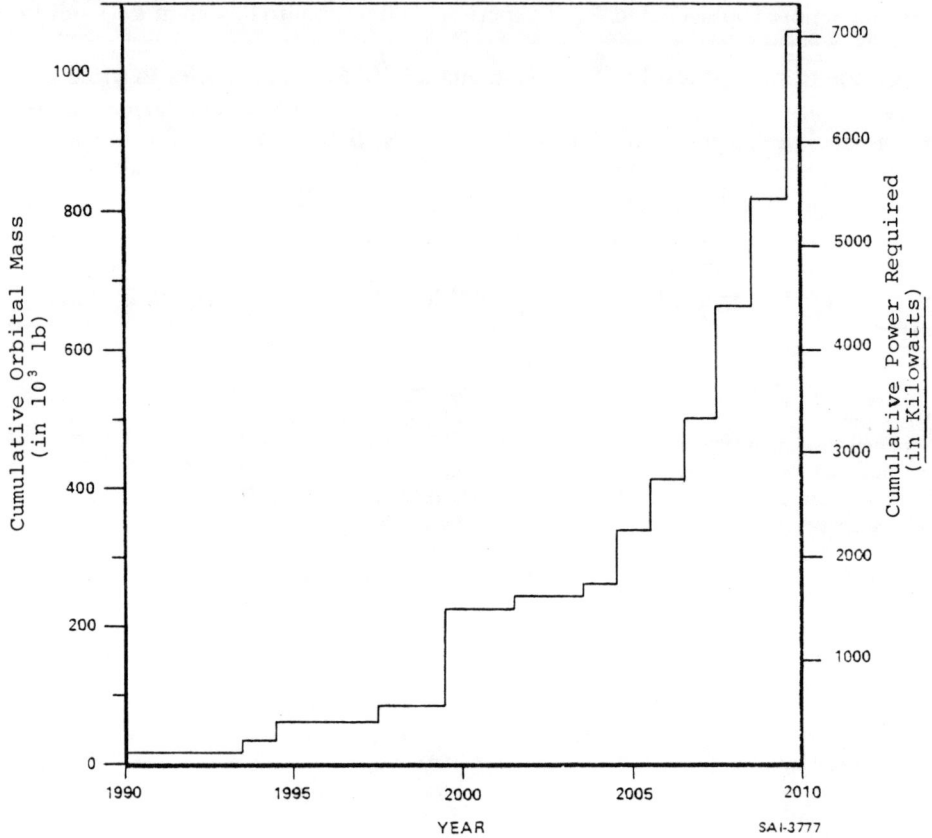

FIGURE 3. One example of the evolutionary mass and power requirements at geosynchronous orbit
generated by selected initiatives.

    C. Solution growth (TGS, growth environment vs. morphology)
    D. Float zone (Marangoni convection, radial segregation, interfacial stability)
2. Metallurgical alloys and processes
    A. Immiscible alloys
    B. Magnetic composites
    C. Metal Foams
    D. High G/R solidification
    E. Solidification at extreme undercooling
3. Composites
    A. Casting of dispersion-strengthened alloys
    B. Solid electrolytes with dispersed alumina
    C. Particle pushing by solidification interfaces
4. Glasses
    A. Glass fining
    B. Laser host glasses
    C. Optical Glasses with unique properties
    D. Metal glasses
5. Chemical processes
    A. Monodisperse latexes (polystyrene microspheres)
    B. Stability of foams and suspensions
    C. Collodial interactions

    D. High temperature properties of reactive materials
    E. Diffusion controlled synthesis
6. Separation sciences
    A. High volume—high resolution electrophoretic cell separation
    B. Protein purification by continuous flow isoelectric focusing
7. Fluid studies
    A. Nonbuoyancy driven convections
    B. Wetting and spreading studies
    C. Role of convection in processes (electrokinetic separation, electroplating, corrosion, etc.)

Additional information on each of these research areas can be obtained from the bibliography or MPS Project Office previously cited.

From the standpoint of anticipating space industrial activities, the current and past research provides the technical base from which extrapolations to possible commercial endeavors can be made. The potential products listed in the previous chapter were arrived at using this extrapolative approach and projections of demand for these items. In the near term, given the high cost of Space Shuttle transportation only, very low volume, high value products can be anticipated. As the cost of obtaining raw materials in orbit decreases, however, more products and larger production runs will become economical. All products listed in Chapter 1 have two things in common: relatively high cost (several hundred to several thousand dollars/pound) and large existing or projected markets. What this may mean to space industry activities in the mid to latter 1990s and beyond will be discussed later. The current activities leading to early commercialization of high value, low volume products in the mid to later 1980s will be the focus here.

The following paragraphs are excerpts from the NASA publication CMPS-200 prepared by the Commercial Materials Processing in Space Office of Marshall Space Flight Center, Alabama. These best summarize what is currently happening and will be developing over the 1980 to 1983 time period.

What is MPS?
MPS, as presently envisioned, encompasses the following:

1. Ground-based research to create a sound scientific basis for investigations in space.
2. Investigations of material properties and material phenomena in the unique environment of space.
3. The making in space of exemplary or model materials to serve as a point of reference for ground-based materials and processes.
4. Applications investigations and feasibility demonstrations of space made or space derived materials and processes. Space made products are expected to be limited to low volume, high dollar value items.

What areas are presently being investigated?
Principal areas of investigations are crystal growth and solidification, metallurgical materials and processes, chemical processes, glasses, composites, fluid studies, and separation sciences.

What makes the space environment unique for materials processing?
Microgravity—the one truly unique aspect of the space environment is microgravity, or "weightlessness" as it is sometimes called. In space, the effect of gravity can be reduced to one millionth (i.e., down to $10^{-6}g$'s) of those experienced on the surface of the Earth. In microgravity, the forces of buoyancy, sedimentation, and thermally driven convection can be virtually eliminated. Other forces and mechanisms such as surface tension and diffusion, the effects of which are often masked on the surface of the Earth by gravity, become dominant.

High Vacuum—at the orbits normally used for space flight, high vacuum can be obtained in very large volumes and with very high pumping rates. Space vacuum is unique in that is can be used in conjuction with microgravity.

What are some of the phenomena which have been observed in the work to date?

1. Compositional homogeneity to a degree not achieved on Earth in materials melted and solidified in microgravity
2. Vapor growth of crystals to a size and degree of perfection not yet accomplished in Earth-based processes
3. Mixing of immiscible materials (i.e., materials that normally do not mix on Earth)—these materials remained in suspension until solidification occurred
4. "Containerless processing" whereby materials were melted and solidified without their being in contact with the walls of a container (in containerless processing, container wall nucleation effects and contamination of the material by the crucible are avoided and independent control of liquid shapes is possible
5. Casting made by using only a thin oxide skin as the mold
6. Improved resolution and purity in the separation of live cells (using electrokinetic separation)
7. Extended, stable floating zones of liquid materials with low surface tension
8. Improved alignment and distribution of constituents in composite materials (for example, continuous fibers in a matrix of a second material)
9. Gas bubbles suspended in a liquid for an extended period of time

What are some of the applications which have been suggested?

Advanced materials research—perhaps the most productive initial application of the unique environment in space will be to further basic knowledge of materials and processes. The low-gravity environment can significantly aid in the investigation of basic material properties. For example, new insight can be gained by studying materials in the absence of natural buoyancy, sedimentation, and thermally driven convection. Solidification processes at a high degree of supercooling can be studied and, using containerless processing, physical properties data on materials that are highly corrosive in the molten state can be obtained. Small samples of advanced or exemplary materials can be made for evaluation. Further, a better understanding of the role played by gravity in a production-process can be gained, which can lead to better methods for controlling the process on the ground.

Experimental investigations of space-unique materials and processes:

Polystyrene microspheres in very uniform sizes—precise, uniform polystyrene microspheres in sizes up to 40μm are potentially very useful in medical diagnostic work and scientific instrument calibration. Sizes of 2 μm and less can be made on Earth and are presently being used. Currently, larger sizes (2 to 40 μm) of adequate uniformity cannot be made on Earth in practical quantities because of sedimentation and creaming problems during the polymerization process. A significant market is expected for larger sizes. The feasibility of making these larger particles is to be demonstrated on early Shuttle flights. Demonstration of commercially viable quantities of the desired quality and sizes is expected by 1982.

Fusion targets—in the search for new energy sources, inertial confinement fusion (ICF) is being investigated by the Department of Energy. One approach to ICF employs precise ultra-thin-wall glass microsphere targets which contain a mixture of deuterium-tritium fuel. The processes by which the spheres are made can be studied in space where specimens can be made and observed in a gravity-free environment. These studies are intended to result in improved manufacturing techniques on the ground.

Electronic materials—several areas in the semiconductor field could benefit from better understanding of segregation, diffusion, kinetics at the liquid-solid interface, and the effect of imperfections. Among these are imaging arrays, infrared and nuclear detectors, and large focal plane arrays. In the microgravity environment of space, better control of disturbing flows, better compositional homogeneity and lower defect densities are possible. Exemplary materials, processed in space, can be used to better understand problems, limitations, and potential improvements in Earth-made semiconductors.

High technology optical materials—the ability to melt and solidify material without its being in contact with container walls offers new opportunities to glass manufacturers to eliminate specimen contamination by the container and to prevent heterogeneous nucleation or devitrification. Since many potentially useful glasses are highly corrosive in the melt and tend to devitrify easily, space processing should lead to important products such as glasses with unique indices of refraction and dispersion for optical components, materials with improved infrared transmission, more efficient laser host materials, Faraday rotators, and low-loss windows for high power laser systems.

New metallic compositions—a fundamental problem in metallurgy is to mix and solidify materials with varying densities and melting points into alloys with a desired microstructure. A low-gravity environment provides new possibilities for controlling the solidification process. In this environment, thermal-gradient-to-growth-rate ratios that cannot be achieved on Earth are possible. It may permit new alloys of immiscible materials, off-eutectic materials, dispersions, and foams to be studied. Improvements in the microstructure of currently produced alloys may also be possible. Applications may include improved turbine blades, permanent magnets, super-conductors, and other high-technology materials.

Separation of biomedical products—there is a continuing need in the biomedical community for better separation and purification techniques for specific cell types, cell components, cell products, and proteins. The low-gravity environment of space offers a new freedom to control unwanted convective flows resulting from variations in fluid densities arising from changes in concentration or temperature. This may result in improvements in throughput and resolution of electrokinetic separation processes such as continuous-flow electrophoresis and continuous-flow isoelectrofocusing. These processes could become important for producing significant quantities of selected cells and proteins for research and perhaps for therapeutic use.

What is NASA doing to facilitate the commercial use of space for materials processing? NASA is

1. Developing policies and guidelines to stimulate and simplify participation by commercial firms and considering new procedures specifically tailored to encourage commercially oriented investigations and demonstrations
2. Continuing to support and conduct scientific studies in areas where potential benefits may be realized
3. Evaluating the need for an inventory of general purpose MPS hardware which would be made available to commercial firms on a use-charge basis
4. Accelerating the flow of information to potential commercial users through direct information exchanges with industrial scientists and managers, publications, and other communicative outlets

Which U.S. Firms or organizations may engage in MPS for commercial purposes? Any firm, institution, or individual may conduct MSP, provided the work is consistent with NASA's objective of fostering public benefits through commercial use of the technology. The organization or individual will be required to furnish NASA with sufficient information to verify peaceful purposes, and to insure Shuttle safety and compliance with applicable laws and regulations.

What avenues are open to commercial firms interested in performing MPS? Involvement by commercial firms may proceed along any one of three basic avenues: ventures which are fully funded by private industry; joint endeavors where each of the parties assume responsibility for a portion of the total effort; and experiments or demonstrations funded directly by the government. As discussed in the succeeding section, invention and data rights vary with the option selected. These options are summarized as follows:

Privately funded ventures—(1) Using privately owned flight hardware—private industry may fund the entire venture, including development of experimental packages, a pro rata share of integration, operating and flight costs, etc. It is understood that all equipment must meet flight safety standards and integration requirements. (2) Using government-owned flight hardware—NASA is considering making government-owned MPS equipment available to commercial firms on a use-charge basis. This will permit firms to conduct experiments and demonstrations while minimizing their capital investment in MPS equipment. The firm will be expected to pay a use charge for the MPS equipment as well as its pro rata share of integration, operation, and transportation costs: NASA will provide technical information on the equipment, limited engineering services, and assistance with NASA procedures such as safety certifications. Terms, conditions, and use charges, now being developed, are scheduled for publication in 1979.

Joint endeavors—NASA is interested in joint endeavors with commercial concerns. In a joint endeavor, each of the parties agrees to be responsible for specific portions of the total venture. In essence, industry can make NASA an offer, stating what the firm will do and what will be expected of NASA. Joint endeavors can be used for a variety of ventures. For example, a joint endeavor might simply call for coordination or collaboration between an industrial scientist and one of the principal investigators involved in NASA sponsored research. In such a case, the industrial scientists, who would be termed a "guest investigator," would participate to an extent mutually acceptable to his firm, the principal investigator, and to NASA. However, a joint endeavor could cover a substantial research and development effort involving major flight experiments and commercial demonstrations. In joint endeavors, NASA does not fund any portion of the work to be performed by the firms.

Government-funded, commercially oriented experiments and demonstrations—periodically, NASA solicits proposals for MPS experiments and demonstrations through formal announcements. Interested parties may receive these announcements upon request. In addition, industrial firms may submit an unsolicited proposal at any time. Proposals selected will be fully or partially funded by NASA, depending on the specifics of the proposal involved. Also, NASA welcomes suggestions on ways to foster more commercial interest in the MPS program and to encourage more commercially oriented investigations. What about rights to inventions and data resulting from these ventures? Rights to inventions and data will depend on the type of venture involved.

Privately funded ventures—when a venture is privately funded, as previously described, NASA will not acquire rights to inventions and data resulting from the venture. In those exceptional instances where the administrator determines that the venture may have significant impact on public health, safety, or welfare, NASA may require, as a condition for flight, certain assurances that the results will be made available to the public on reasonable terms and conditions. Should it be necessary for NASA to have access to privately developed trade secrets to carry out the venture, an agreement for protective handling will be developed.

Joint endeavors—in joint endeavors where NASA does not fund any portion of the work done by the firm, the allocation of rights to resulting inventions and data is subject to negotiations between NASA and the firm. NASA's objective is to stimulate public benefit through private initiatives. However, the agency would normally expect to receive appropriate consideration for its part in the endeavor. For example, where NASA's participation is substantial, the government may seek a royalty-free license in resulting inventions and data for U.S. governmental purposes only. However, considerations can vary from endeavor to endeavor, depending on the specific circumstances involved.

Government-funded experiments and demonstrations—where NASA directly funds experiments and demonstrations, either fully or in part, the firm will be working under contract with NASA and, as such, the agency is required by law to take title to inventions made in the performance of the contract work. The administrator of NASA may, however, waive title to such inventions to the firms where it is determined to be in the interest of the U.S. government to do so. Petition for waiver of title must be made in accordance with NASA's Patent Waiver Regulation (14 CFR 1245.1). In applying the Regulation, both the need for incentives to draw forth private initiatives and the need to promote healthy competition are weighed. In general, it is NASA's objective to promote early utilization, expeditious development, and continued availability of new technology for commercial purposes and public benefit. As to data, NASA normally acquires, on behalf of the U.S. government, rights to data first produced under such contracts.

When can commercial firms begin conducting MPS experiments and demonstrations? At the present time, a limited number of short-duration, low-gravity experiments can be conducted using NASA's drop tower, aircraft flying a "low-gravity" profile (parabolic trajectory), or suborbital rocket flights. Also, limited opportunities are now open for industrial "guest investigators" to participate in ongoing, NASA-sponsored, ground-based and flight MPS research. With regard to space flights, according to present plans, a few opportunities will be available on the Space Shuttle in 1982 or 1983. Within 5 years, NASA expects MPS flights to be routinely available to commercial firms.

What other countries are sponsoring work in materials processing in space? The Federal Republic of Germany is active in MPS research, has flown several experiments, and is scheduled to fly additional experiments on early Space Shuttle flights. The German program includes industrial participation. The Soviet Union flew several material experiments on the Apollo-Soyuz mission in 1975. Reports indicate that additional MPS experiments in a variety of fields are being conducted by the Soviet cosmonauts in manned Salyut space laboratories. Organizations in such countries as Australia, Belgium, Canada, Denmark, France, Italy, Japan, Spain, Sweden, and the United Kingdom are presently active in MPS research.

IV. IN-ORBIT SERVICES

A set of space industry activities of interest in the early to mid-1990s will be services in Earth orbit. These will include providing platform space, power, maintenance, servicing, and repair. The market for these activities will exist in both low orbital and high orbital regimes.

The low orbit customers will be those involved in manufacturing operations of a small to moderate scale not compatible with the shuttle or its derivatives. All or part of the above-mentioned support could be applied to these operations.

The high orbit customers will be the owners and/or operators of the expanding information services systems. The mass and power requirements mentioned earlier coupled with the complexity of these satellites will make highly reliable, long-lived throwaway systems unattractive. Other factors such as geosynchronous orbit crowding and the accumulation of space junk will also force the need for integrated platforms, central power and access for service, repair, and update. Initially these things probaby will be handled by remote manipulators, but human presence will almost surely be required by the mid-1990s. This implies the need for space transportation systems beyond the shuttle capability to transfer from low Earth orbit to high Earth orbit. These are discussed in a later chapter.

Also implied, in the longer term (latter 1990s and beyond), is the probable need for a high orbital work force. Earlier operations would be envisioned as sorties to do a series of tasks in a short time and return to Earth. Reference 3 has shown, however, that the scale of activity in space with or without the Solar Power Satellite (SPS) program described in another chapter will probably be sufficient to justify a permanent work force in high orbit. The number of people required will of course be heavily influenced by the existence or nonexistence of a program the size of SPS.

V. POTENTIAL FOR USE OF NONTERRESTRIAL MATERIALS

Since the publication of O'Neill's[5] classic paper (1974) on lunar materials (LM) utilization for construction in Earth orbit, the threshold of practicality for such use has been at issue. Such a threshold had been previously examined by both Ehricke and researchers at the NASA/Ames Research Center regarding the supply of raw materials for the Earth. More recently, comparisons have been done between very large-scale programs involving SPS construction in space.[2,3] For Earth use of raw materials it was generally concluded that a lunar-based industry could not compete with an Earth based except in a situation of great need and great scarcity. Studies on SPS have shown thresholds of 1 to 30 units on cost equivalency between terrestrial and nonterrestrial materials. Orbital activities on a scale significantly smaller than SPS have not previously been examined in a systematic way.

The practicality of using lunar materials in space depends on several factors: the technology of obtaining and converting the raw material; the potential existence of a market sufficiently large to warrant the investment; and the capability to compete favorably against an Earth-based industry. Sufficient study, experimentation, and exploration have been accomplished to establish that the necessary technological goals are achievable. Practicality thus depends on the projected markets and potential for competitiveness. These are the areas addressed in a recent, brief study effort, the results of which are presented in Reference 2.

A. Projected Earth Orbit Materials Market

To evaluate the application of NTM to these potential product and service areas, data as presented in this and the preceding chapter were compiled. Projections of total mass vs. time and location in Earth orbit were compiled and an evaluation of the percentage of this total requirement deemed compatible with lunar materials was made. The projections were constrained to the period of 1990 to 2010. Figure 4 presents the total demand projected as a function of market assumptions. Note that although GeoSat Information revenues were several times those of materials in Figure 2, the mass requirements are dominated by the products projections.

There are two reasons for the noted difference. First, the GeoSat mass is in finished, functioning product, obviously of higher value than the raw material being supplied a processing plant in Earth orbit. Secondly, pound for pound the GeoSat represents a significantly higher intrinsic revenue potential over its life span. Conversely, one would be willing to spend significantly more to obtain finished GeoSat parts in high Earth orbit (HEO) than to obtain unprocessed material in low Earth orbit (LEO).

The data of Figure 4 are presented in Figure 5 varying with time between 1990 and 2010. The apparent Sigmoidal behavior of the curves is driven by the use of classical market penetration and capture assumptions. The upturn after about 2006 is driven by the rising dominance of GeoSat markets not projected to saturate by 2010. Historically, the opening of a new economic operating regime has resulted in growth more along the dashed line path shown in Figure 5. The limitations inherent to the theory and assumptions underlaying the three projections shown make prediction of the historical type curve impossible. A broader base of product and market characterization would be required to begin to approach such a projection. Although it may be possible to gather such data it has not been accomplished to present. For purposes of the present study the three projections were sufficient to compare Earth- vs. lunar-based material supply.

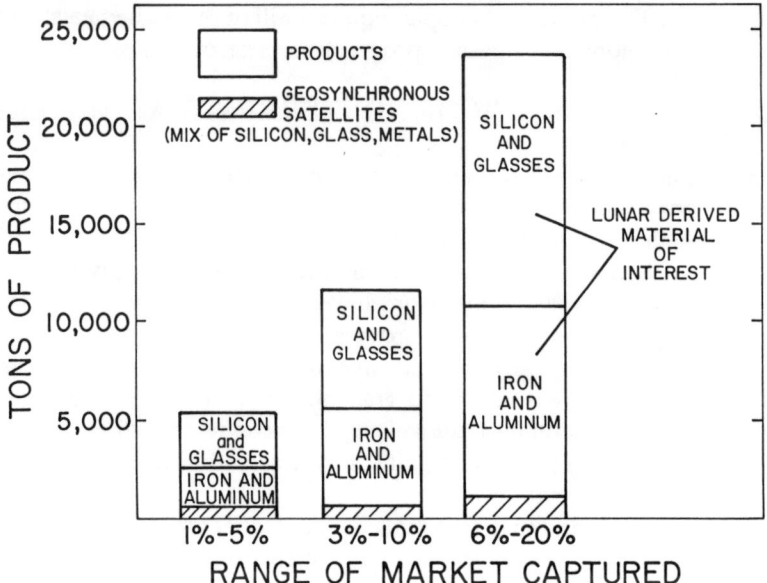

FIGURE 4.   Possible demand for lunar materials based on level of market satisfied
by space utilization between 1990 and 2010.

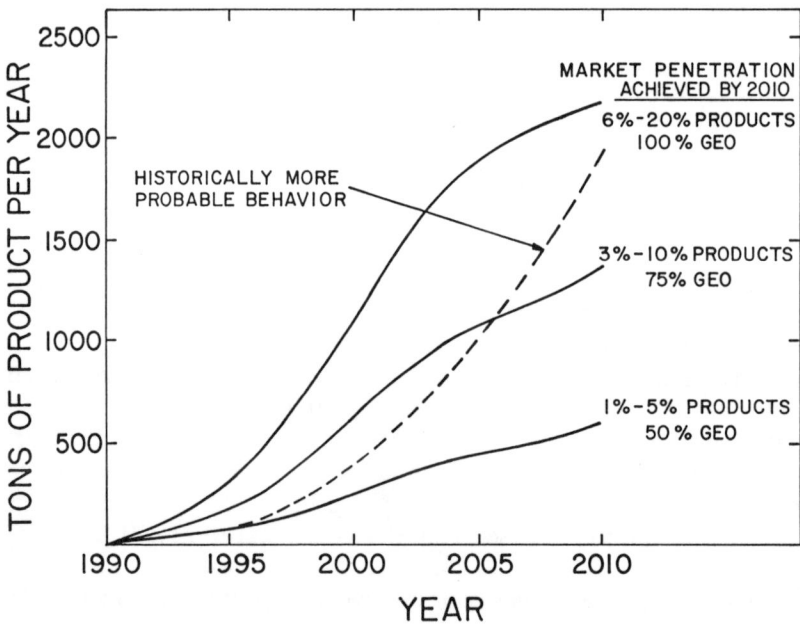

FIGURE 5.   Possible demand for lunar materials based on proportion of specific
commercial markets captured by space based products and services.

B. Cost of Materials Transport from Earth

The total cost to provide transportation to LEO and GEO at any point in time for the
three scenarios of Figure 5 is dependent on three things. First, the Design,
Development, Test and Engineering (DDT & E) cost to provide a capability such as
increased mass per launch or reduced cost per ton of payload in orbit. Second, the cost

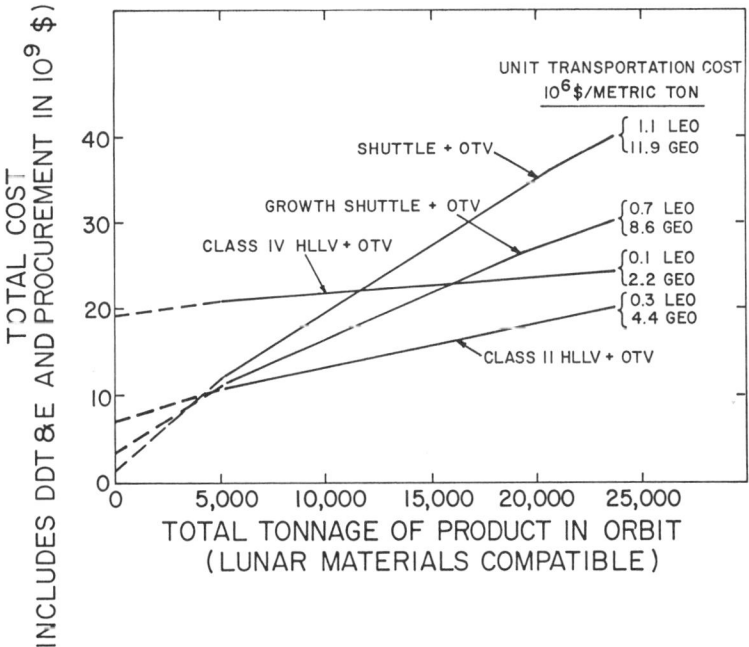

FIGURE 6. Transportation cost as a function of total required tonnage in the market scenarios.

of procuring sufficient vehicles to handle the traffic during the peak launch year. Third, the operations, expendables, and refurbishment costs which occur due to launch of payloads.

A national decision to embark on the usually expensive and rather risk-laden venture of launch system development is typically driven by a combination of economic tradeoffs and political perceptions. In the development of space industry the economic driver will be dominant over all but technological considerations. No foreseeable single program or set of government programs will provide justification for a successor to the Space Shuttle until the 1990s or beyond.

However, since this study addressed the practicality of establishing a *single* materials industry in space (perhaps composed of several elements) it was appropriate to examine a range of vehicle combinations with potentially lower operating costs than the Shuttle and Shuttle/Inertial Upper Stage (IUS). The vehicles considered were **Earth Launch**—(1) Basic Space Shuttle (65,000 lb LEO); (2) Growth Shuttle (100,000 lb LEO); (3) Class II Heavy Lift Launch Vehicle (HLLV, 350,000 lb LEO); (4) Class IV HHLV (600,000 lb LEO); and **Orbit Transfer**—Fully Reusable Orbit Transfer Vehicle. The cost for the total transportation requirements of the three 20-year programs outlined above were computed using the unit transport costs shown on Figure 6. These values were arrived at by consideration of such parameters as true load factor (gross capability minus shipping container for example for raw materials).

The results of the total cost evaluations are presented on Figure 6 for each vehicle combination presented. Although a nonlinearity is surely present, the data at 5420 tons (the minimum program) are connected to the DDT & E costs by a straight dashed line to estimate cost in that regime.

C. Lunar Materials Scenarios

Making the comparison between Earth-based and lunar-derived materials required development of an operational scenario for obtaining, transporting and processing the raw materials. Time and resources did not allow a detailed evolutionary scenario to be developed for all three demand curves of Figure 5. Some tradeoffs were made at the lower total demand level and assumed to apply in the other two cases. A specific example was the evaluation of raw material transport modes. Since the fuel (liquid hydrogen) for a chemical system was assumed to come entirely from Earth, it was advantageous to use a small Mass Driver system as defined by O'Neill[5] to project ore stock from the moon to a "catcher". Surprisingly, this proved advantageous even at the lowest demand level examined (5420 tons of products over 20 years). Also, as has happened repeatedly in other studies, the placement and operation of the chemical processing plant in space vs. on the lunar surface again seemed advantageous. This was partially driven by the real value assigned "slag" based on its usefulness as radiation shielding.

The capability requirements of the individual components of the operational system were determined during initial phases and the first installations sized accordingly. For example, the materials process plant in the low scenarios was sized to be in full operation by about 2010, operating intermittently or at reduced capacity from 1990 through 2009. Capacity was incremented according to Figure 5 beyond 2000. The remaining two demand curves were satisfied similarly using the basic modular unit of the low scenario.

The lunar operations were sized based on a Mass Driver capable of meeting peak year demands and operating intermittently or at reduced capacity during prior years. Inefficiency in sizing for the early years is probably compensated by the reduced cost of operations since no people require support during extended shut downs.

The manufacturing plant is not required by GeoSat demand until about 1998. The production rate grows rather rapidly after that, however (see Figure 3). A modular system was again assumed with a factor of four growth by 2007.

For transportation, lunar growth oxygen extracted on the surface of the moon and at the processing plant was used extensively for both chemical and electric propulsion. No Earth launch system beyond the presently defined space shuttle was assumed through the entire operations period. All hydrogen, nitrogen and process plant make-up reagents were transported from Earth. All food and hydrogen to make water for all crews was also assumed hauled from Earth although a simple water recovery system was assumed.

Thus the following represent the elements of the space materials and component production facility: (1) transportation—Space shuttle to LEO, LOX/H_2 POTV for passenger orbital transport, LOX/H_2 LTV for lunar surface transport, LOX COTV for hauling cargo between orbits, Mass driver to eject ores from the Moon; and (2) Facilities—Lunar mining and bagging, Lunar habitat, Mass catcher, Chemical process plant, Manufacturing plant (post—1998), Space habitat, LOX plant and depot (lunar surface and orbit). (Traffic levels for the three demand scenarios do not warrant way stations at LEO or lunar orbit.)

The initiation and build-up scenario for the low demand projection proceeds as follows. (1) Twelve shuttle flights lift 300 tons to LEO over a few months span. Unmanned cargo flights transport 35 tons to the selected lunar mining site. Two crew modules of eight personnel each land at site. (2) Crew of 16 assembles and activates mining operation over 60-day period (Mass Driver pacing item). Bury two crew modules for extended use. All structure and systems designed for maximum reuse or cannibalization. Once system is operational, 12 return to Earth and regular crew of 4 remain to operate and maintain automated system for 6 months. (3) Capture MD output and transport to processing plant. (4) Shut down lunar operation for 5 yr and run

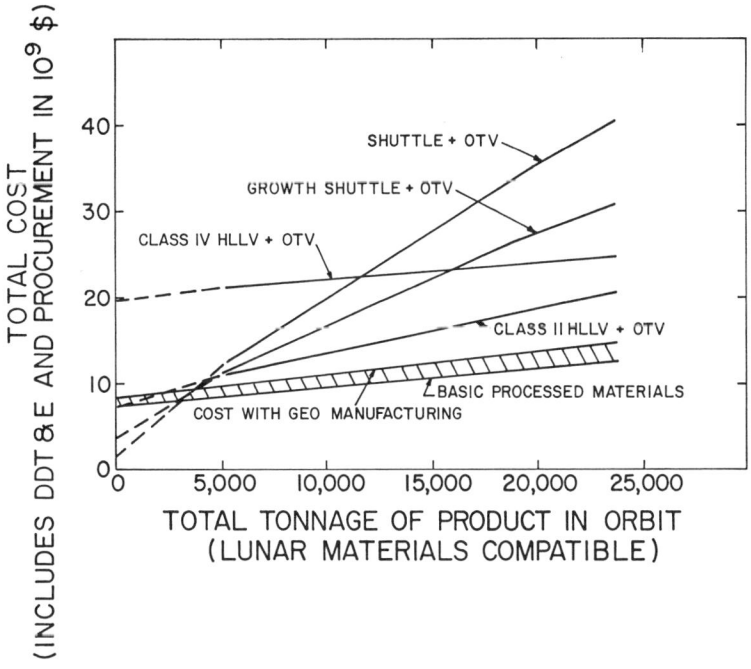

FIGURE 7. Comparison of transportation cost options to the lunar materials option.

system off initial accumulation or operate the mine for short periods annually. (5) Parallel to 1 and 2 deploy a processing plant capable of ultimately achieving 600T/year output products. The unit would weigh about 10 tons with a 10 ton, 500 kW power supply. A crew of six would operate and maintain the system during intermittent operating periods from 1990 to 2001. After 2001 the crew would be full time. (The ultimate capacity of the other two scenarios require a 45 ton, 1.5 MW and a 90 ton, 3 MW plant with crews of 8 and 10, respectively.) About two shuttle flights required to deploy. (6) Rotate all crews on a six month basis and use a O-g habitat at the processing/manufacturing habitat. (7) Deploy a 40T/year output plant of about 17t mass in 1997. Begin full scale operations in 1998 to 1999 time frame. Total capability is incremented in 40T/year output increments to about 160T/year in 2010. Six people are required for the basic unit operating full time; 12 for the 2010 operation level. The mid and upper scenarios used 50T/year and 80T/year capacity plants as the fundamental module.

D. Cost Practicality Thresholds

The cost for transportation from Earth of the total demand tonnage is compared to the cost of utilizing lunar materials in Figure 7. On the basis of simple integrated cost over the 20-year span in constant 1979 dollars the lunar option is lowest for all three scenarios examined. Cross-overs occur at less than 4000 for the Shuttle-based options.

Figure 7 does not address the economic issues associated with the various options for material supply. When such considerations are made, the initial investment for both the lunar option and the Class II HLLV + OTV will drive the integrated cost curves upward. Such an economic examination should not change the basic ordering of the results shown, considering the spread of values.

In dealing with any overall economic assessment, however, other factors should be taken into consideration. For example, the growth of space industry to the levels implied here for post-2000 may not occur unless a more economic mechanism than Earth launch for material supply is found. The minimal cost per pound for lunar material has not been identified in this study; only a rather simple set of comparisons made. The levels of activities envisioned for SI without SPS will never justify the very large Class IV plus vehicles potentially capable of less than 0.11×10^6 $/t ($50/lb) to Earth orbit. Unit costs below this threshold appear feasible with an efficient system to obtain lunar materials at these lower quantities. Very large scale Materials Space Processing may depend upon achieving such goals.

The cost of operations at geosynchronous orbit for the large Information initiatives must be considered in an economic assessment also. A lunar utilization industry provides the basics for such an operation by definition, whereas the habitat DDT & E and operational capability costs must be added to the transportation options shown on Figure 7. Also, the existence of a good supply of very cheap shielding material is not inconsequential when considering operational modes for manned geosynchronous operations.

E. Key Technologies

A great deal of work needs to be done on several technologies which directly affect cost and efficiency for obtaining and using lunar materials. A few of these are compiled in the following paragraphs.

1. Transportation

(1) POTV—develop a capability to transfer personnel and payloads to geo. and lunar orbit and return. All the chemical propulsion and life support technologies are basically state of the art. (2) COTV—develop the technology and demonstrate the feasibility of using oxygen as a propellant in a high efficiency, high reliability ion drive engine. Some fundamental research and demonstration work can be done relatively inexpensively to seek out and solve problems pursuant to building scaled-up hardware. (3) LTV—begin tradeoffs and preliminary design studies related to defining a minimum cost LTV suitable for passenger transport to and from the Moon with extended, untended loiter time in lunar orbit. Two options should be studied: a passenger only and a passenger plus cargo *down,* passenger up. Examination of inexpensively hard landed or semisoft landed supplies should be examined.[6] (4) Mass driver—continue the development work at Princeton leading to detailed designs; operation under simulated lunar conditions; maximized efficiency for mass; optimal modularization; etc. Such ancillary issues as techniques for long term containment and maintenance of liquid and gaseous helium should be identified early to enable suitable research efforts to begin. Demonstrate the minimal manpower required for deployment of a Mass Driver.

2. Facilities

(1) Lunar mining and bagging—begin preliminary design studies on a minimal, modularized, and high automated facility. Design and do scale demonstration of equipment and techniques for handling lunar materials (particularly fines). Design and do scale demonstration on the manufacture of fiber glass and woven bags. Design an automated system suitable for MD loading at various rates. Design a surface mining system capable of handling the fines and rubble. (2) Lunar habitat—design a minimal habitat suitable for short-term (60 days) occupancy by 8 to 10 people and long-term (6 months) occupancy by 2 to 4 people. Should be transportable from Earth orbit as a single unit and "suitable for burial" on the lunar surface for protection. (3) Mass catcher—design a minimal system capable of intercepting and transporting or handing

over 10 to 100 tons of bagged lunar material. The key technological needs of such a system should result from this preliminary design so that research efforts may commence where necessary. (4) Chemical process plant—demonstrate the capability in the laboratory to extract oxygen and desired metals from simulated lunar material using the three or four most promising techniques previously studied. Using criteria developed during these demonstrations, select one route and do a detailed plant design to obtain more insight into technology needs. Phase research accordingly. (5) Manufacturing plant—major rapid development of geo based services may occur than predicted in this study. Studies to identify specific manufacturing machine needs leading to preliminary design of such machines should be conducted. (6) Space habitat—design a minimal suitable O-g habitat for 6-10 people capable of growth and shuttle compatible. It should be designed with the use intended here in mind. Previous detailed studies for LEO space stations of similar size may provide a good starting point. (7) LOX plant and depot—preliminary design of a plant and depot along with research and demonstration of techniques for extracting O_2 from lunar soil. Design of a light weight, automated plant complete with raw material gathering, liquefaction and storage capacity is required.

3. Site Selection

(1) Lunar—detailed study of the lunar surface and trajectories for ejected bags form the minimal basis for mining and mass driver site selection. An automated inspection capability ultimately may be required. (2) Space—the optimal siting of the materials processing and manufacturing plants should be evaluated as a function of costs associated with all transport functions.

4. Market Development

(1) Materials and products—extensive ground-based study and research and space-based experimentation and demonstration will be required to develop the unique products of sufficiently high value to warrant building plants of the scale discussed here. An aggressive set of programs directed toward developing a set of such proudcts compatible with lunar materials is required. Obviously experimental work involving silicon, silica, glasses, aluminum, iron, and steel would be of value. The use of inexpensive "Get-a-Way" specials can become a key to this work. (2) Satellites—develop the sub-set technology areas of large power systems (50kW to 10MW), large structures, control and antenna design. Work in these areas is in progress and should be promoted aggressively in the 1980s to assure industry investment and operational capability development in the 1990s.

5. Near-Term Programs

(1) Power systems—NASA and industry should actively promote development of free-flying power systems begining with the 25-50kW module currently being considered. A continuing program leading to development of systems with continuous output on the order of 1 to 10 MW should be pursued. (2) Large space structures—research and development in efficient fastening and assembly, control, and material sciences (especially long term exposure response) should be aggressively pursued in the 1980s. (3) Orbit transfer vehicle—a manned geosynchronous orbit capability is the next major step in extending our reach into space. The system should be designed and developed with the lunar mission discussed in this paper as part of the criteria. Availability by 1988 to 1990 would seem most appropriate. (4) Solar Electric Propulsion System (SEPS)—develop and utilize a SEPS capability using 25 to 50kW as a minimum during 1980s. Pursue program expansion to larger power systems and

develop oxygen propellant technology. (5) MPS—expand the current program to allow involvement of greater numbers of experimenters and paylods in the early 1980s. Steadily promote the increased involvement of industry in joint endeavors with government to make commercial activities feasible. (6) Communications satellite technology—promote a coherent technology base suitable for providing the necessary state-of-the-art in the 1990s leading to large scale, high power systems.

VI. EVOLUTIONARY SCENARIOS

The development of an industrial base in space similar to the scenarios discussed previously will depend strongly on cooperative arrangements worked out between elements of government and industry. These arrangements will be particularly important during the early to mid-1980s when risks, uncertainties, and pay-back periods make private investment almost impossible. Current NASA and Congressional efforts to promote commercial applications of the Space Shuttle are a strong, positive step in the direction of meaningful joint endeavors.

If risk-sharing arrangements on research and development can be arrived at, it seems quite likely that the fundamental markets will begin to develop. At that point sufficient knowledge will be in hand to encourage multiple industries to make investments against reasonable use guarantees. That is, user A assures, guarantees, or warrantees supplier B that his product will be used at some minimal price and quantity. Of course, the first user is ultimately the consumer of goods or services, and careful, in-depth market research work is an absolute must. However, once the market is sufficiently characterized, the proper agreements can begin to fall into place through the ultimate supplier of raw material. Of course, if the necessary investment in new research and development and equipment can be shown to be sufficiently small, one corporation and/or investor might be attracted to establishing the entire set of systems and facilities. It seems more likely that the development of individual elements of the industrial system will evolve similar to the more segmented historical Earth based approach which distributes risk during initial phases.

Although this study examined the period 1990 to 2010 based on previously existing data, it unfortunately appears unlikely that a lunar materials base industry can evolve rapidly enough to be in operation by 1990. The constraints are not technological as related to development of techniques, hardware, or software. They are technological in the sense that the true potential for generating high value products and large communications systems that have markets must be determined. Thus the results of MPS experiments and market development in the 1980 to 1990 time period will be crucial as will the large GeoSat initiatives. The earlier a uniqueness is established for certain products and the market determined for large satellites, the sooner will begin the drive to obtain the products and services at the least possible cost.

REFERENCES

1. Lunar Resources Utilization for Space Construction—Final Report, NASA Contract NAS9-15560, General Dynamics/Convair Division, San Diego, Tex., 1979.
2. **Driggers, G. W.,** Is lunar material use practical in a non-SPS scenario?, in *Proc. 4th Princeton/AIAA Conference on Space Manufacturing Facilities* Gray, J., Ed., (in press), May 1979.
3. Space Industrialization Study—Final Report, NASA Contract NAS8-32197, Science Applications, Inc., April 1978.
4. *Space Industrialization—Final Report,* NASA Contract NAS8-32198, Rockwell International/Space Division, April 1978.
5. **O'Neill, G. K.,** The colonization of space, *Phys. Today,* 32, 1974.
6. **Clark, A. C. and Smith, R. A.,** *The Exploration of The Moon,* Harper and Brothers, New York, 1954.

Chapter 3

SOLAR POWER SATELLITES*

Peter E. Glaser

TABLE OF CONTENTS

* This chapter was first published as "The Development of Solar Power from Satellites" for *Advances in Energy Systems and Technology,* Vol. 2, Academic Press, New York, 1979.

* This chapter was first published as "The Development of Solar Power from Satellites" for *Advances in Energy Systems and Technology,* Vol. 2, Academic Press, New York, 1979.

I. INTRODUCTION

Energy has been the key to the social development of man and an essential component in improving the quality of life beyond the basic activities necessary for survival. Consequently, conversion of various energy resources into power has been and will continue to be essential. But the amount of energy and the changes in the mix of resources used to generate power is dictated by technological, economic, environmental, and societal considerations. Therefore, it has been recognized that no one energy source, by itself, is likely to meet all future power needs. Inasmuch as there are many uncertainties inherent in achieving the potential of the various alternative energy conversion methods, the possible role of solar energy is being reexamined, since it has been recognized that solar energy is the most widely distributed energy source and one capable of meeting a significant portion of future energy demand.

In response to the energy needs of the Industrial Revolution, efforts to harness solar energy accelerated during the last half of the nineteenth century and the beginning of the twentieth century. These efforts subsided with the successful development of energy economies based at first on coal and subsequently on the use of liquid petroleum fuels. It was not until the early 1970s that the development of solar energy conversion methods—by concentrating solar radiation to generate high temperatures to power heat engines, by direct conversion of solar radiation with photovoltaic processes, by photochemical conversion to produce fuels, and by indirect methods such as biomass conversion based on the use of products of photosynthesis, wind energy conversion, and ocean thermal energy conversion—was recognized as a promising alternative to conventional power generation methods.

But the degree to which the application of these solar energy conversion methods will be successful will to a large extent depend on their economic feasibility, which in turn depends on technological advances to reduce their costs and also on the reduced availability and resulting increased cost of fossil fuels. Although solar energy is a widely distributed rsource, the cost of the equipment required to convert it makes it a challenging task to find and introduce the best methods for converting it efficiently and economically into useful forms on a scale large enough to have a significant impact. This task will require considerable effort if it is to strike the appropriate balance among technology, the environment, and society's needs.

Solar energy research and development currently is being directed toward a search for new technology and approaches that can reduce the cost of collection and conversion and toward materials, designs, and processes that will permit low-cost mass production. But such results are unlikely to be achieved quickly—not because of the difficulty of the technology, but because of the limited experience with such technology, and its related socioeconomic factors and institutional constraints, as a result of the—until recently—lack of appreciation of the potential of solar energy as an alternative energy source.

During the 1960s the space program had a most profound influence on technological advances. It demonstrated that evolutionary progress need not be confined to the surface of the Earth. For example, satellites for Earth observations and for communications already had significantly affected the lives of the population of the Earth and the indications were that there was no limit to the uses of space technology for the benefit of society. Therefore, a logical extension of the efforts to harness the sun was to use the developing space technology to overcome terrestrial obstacles, such as inclement weather and the diurnal cycle, to the large-scale conversion and application of solar energy. If satellites could be used for communications and for Earth observations, then it was also logical to develop satellites that could convert solar energy and place them in Earth orbits, particularly geosynchronous orbits (GEO), where they could generate power continuously during most of the year. With their year-round-conversion capability, such satellites could overcome another of the major obstacles to large-scale installations on Earth, i.e., extensive conversion area and energy storage requirements. Thus the demonstrated capability of industrialized society to develop high technology could be applied to the development of solar energy conversion methods in space on a scale that would not likely be possible on Earth.

In the 1960s, the soundness of using the synergism of solar energy conversion technology and space technology led to the concept of the solar power satellite (SPS) to beam power from space to Earth.[16] As conceived, the SPS would convert solar energy into electricity and feed it to microwave generators forming part of a planar, phased-array transmitting antenna. The antenna would precisely direct a microwave beam of very low power density to one or more receiving antennas at desired locations on Earth. At the receiving antennas, the microwave energy would be safely and efficiently reconverted to electricity and then transmitted to users. An SPS system would comprise a number of satellites in GEO, each beaming power to its receiving antennas. Successful development of the SPS would not only provide a global option for power generation on Earth[43] but would remove the limits to growth imposed by nonrenewable terrestrial energy resources. It also could herald an era of international cooperation to tap energy from space and for the development of extraterrestrial resources in the future.

II. AVAILABILITY OF SOLAR ENERGY IN SPACE

Among the primary arguments for solar energy conversion in space for use on Earth is the nearly constant availability of solar radiation in GEO as compared with solar radiation received on Earth. Whether solar energy conversion is accomplished in GEO or on Earth, seasonal changes in the solar constant, depending on the location of the Earth as it orbits the sun, range from a low of 1.309 kW/m² on July 4 to a high of 1.399 kW/m² on January 3. An even lesser effect will result from the variation caused by the formation of sun spots during periods of increased solar activity.

Solar radiation in GEO will be interrupted by eclipses of the Earth of the Sun from 22 days before to 22 days after the equinoxes for a maximum period of 72 min a day. With many SPSs in GEO and depending on orbit position, neighboring satellites might shadow one another for about 15 min. at 6:00 a.m. and 6:00 p.m. during these equinox periods. In addition, a satellite in GEO will be eclipsed infrequently by the moon for as many as 90 min.

The equinox eclipses of the SPS occur when the Earth as seen from GEO is near local midnight. Overall, eclipses will reduce the solar energy received by the SPS in GEO by only about 1% of the total available to it during a year. Because the sun is not a point source (solar radiation arrives on Earth within an arc of 32'), the SPS will not cast a shadow on Earth.

The variations of solar insolation with time and location and the variations of observed values over several decades make it difficult to predict accurately components of solar insolation on Earth. For example, variations in upper atmosphere turbidity during the past several decades have led to differences in direct normal insolation on Earth of about 7%. The observed changes in the annual average solar radiation are of the order of 10 to 20%, as are the anticipated changes in direct normal radiation.[44] The variations in the magnitude of the components of solar radiation caused by daily fluctuations in atmospheric conditions, such as those indicated by rapidly moving small clouds, by long-term changes, such as those caused by introduction of dust into the stratosphere, or by climatic changes that may affect the average yearly cloudiness for several decades could have a significant effect on the performance of terrestrial solar energy conversion systems.

The advantages of locating a solar energy conversion system in GEO compared to the terrestrial locations indicated in Table 1A are apparent. As Table 1B shows, even a hypothetical comparison of the SPS with an idealized photovoltaic system (no storage and retrieval losses), which is assumed to occupy either the active area of the receiving antenna or the total area of the receiving antenna site, favors the SPS by at least a factor of 2 for average weather conditions in a geographically favorable location.

III. TECHNOLOGY OPTIONS FOR SOLAR ENERGY CONVERSION IN SPACE

In recognition of the potential of the SPS as a large-scale global method of supplying power to the Earth, the challenges posed by the SPS concept are being explored through feasibility studies of the technical, economic, environmental, social, and international issues by the U.S. Department of Energy and NASA.[40] In the following sections the status of the SPS development to date is reviewed and the issues that require resolution are highlighted.

A. Photovoltaic Design

As originally conceived[17], an SPS can utilize current approaches to solar energy conversion, e.g., photovoltaic and thermal-electric, and others likely to be developed in the future. Among these conversion processes, photovoltaic conversion represents a useful starting point because solar cells are already in wide use in satellites. An added incentive is the substantial progress being made in the development of low-cost, reliable photovoltaic systems and the increasing confidence in the capabilities of achieving the required production volumes.[39] Because the photovoltaic process is passive, it could reduce maintenance requirements and achieve at least a 30-year operating lifetime for an SPS. Micrometeoroid impacts are projected to degrade 1% of the solar cell array area over a 30-year exposure period. Because of the small probability of occurrence of larger meteoroids, they are less likely to affect the solar cell array.

Several photovoltaic energy conversion configurations applicable to the SPS concept have been evaluated (Figure 1). Two alternative energy conversion systems—one employing silicon and the other gallium arsenide solar cells have been selected as an SPS

Table 1
COMPARISON OF SOLAR
ENERGY AVAILABLE IN GEO
AND ON EARTH

A. Ratios of Solar Energy Available in GEO and in Terrestrial Locations

	Weather conditions	
Typical location	Clear	Average
Albuquerque	3.8	5.8
Buffalo	6.5	16.8
Huron	4.8	8.2
Little Rock	4.9	9.1

B. Ratios of Power Delivered by the SPS and by a Terrestrial Photovoltaic System

	Receiving Antenna			
	Active area weather conditions		Total site area weather conditions	
Typical location	Clear	Average	Clear	Average
Albuquerque	1.72	2.64	1.31	2.02
Buffalo	2.94	7.65	2.25	5.85
Huron	2.19	3.73	1.67	2.85
Little Rock	2.22	4.13	1.70	3.16

Courtesy of Arthur D. Little, Inc., Cambridge, Mass., 1981.

reference system.[33] The purpose of the SPS reference system is to document an SPS configuration for purposes of economic and environmental assessments, and for comparison with alternative technical approaches. The SPS reference system has an output of 5 GW at the receiving antenna connection to the utility system.

The overall performance of the SPS can be significantly affected by orbital perturbations due to the gravitational potential of the sun and moon, solar radiation pressure, microwave radiation recoil pressure, the ellipticity of the equatorial plane of the Earth, rotary joint friction torques, magnetic field interactions, and aerodynamic drag unless appropriate maneuver corrections are applied by thrusters. Inclination effects (north-south drift) require annual corrections. Solar radiation forces affecting orbital periods require continuous nulling. To achieve the desired thrust levels to control drift in altitude and inclination, ion thrusters with argon as the propellant will be used. About 30,000 kg of argon per year will be required. The placement and reliability of the ion propulsion units will be critical determinants of maintenance costs. Methods of achieving control during SPS construction and maintaining operational control of the SPS during eclipses will also have to be developed.

1. Silicon Solar Cells

Over the past 25 years, substantial experience has been gained in the development

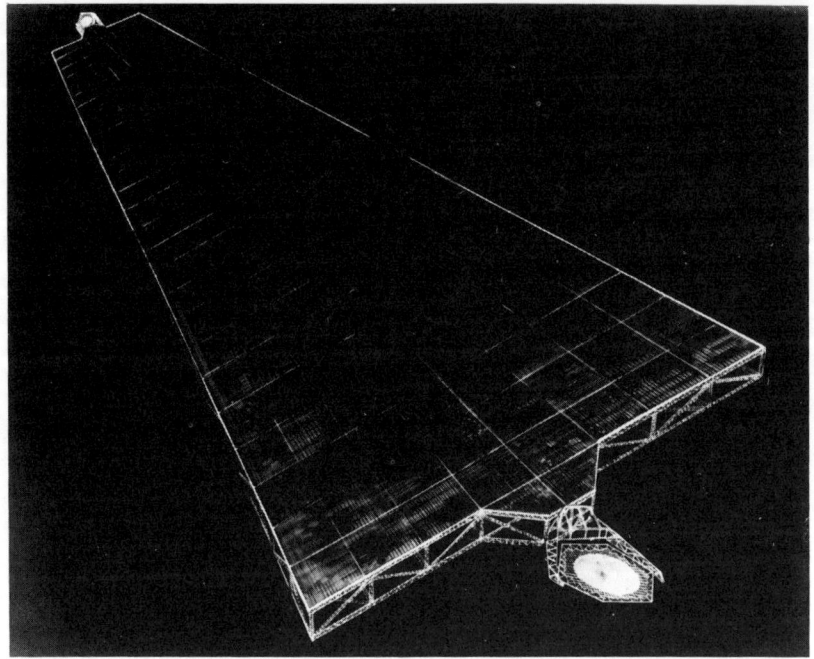

FIGURE 1. (A) 10-GW SPS silicon solar cells. (From Boeing Aerospace Co., Seattle, Wash., December 1977.)[45] (B) 5-GW SPS gallium arsenide solar cells. (From Rockwell International, Los Angeles April 1978.)[35] (C) 5-GW reference configuration-gallium arsenide (From NASA, Lyndon B. Johnson Space Center, Houston 1977.)[3,35]

and production of single-crystal silicon and the solar cells produced therefrom: by comparison, the state of the art of other photovoltaic materials lags. Furthermore, the objective of the U.S. Department of Energy's Photovoltaic Conversion Program is to develop low-cost reliable photovoltaic systems and to stimulate the creation of a viable industrial and commercial capability for mass production at a predictable and reasonable cost. Although the focus of this program is on terrestrial applications of solar cells for widespread use, the successful development of 50-μm single-crystal silicon solar cells and their production in a pilot process with demonstrated efficiencies of 14% at air mass zero indicate that significant progress is being achieved in producing lightweight silicon solar cells.[27]

The SPS solar cells will have to be highly efficient, of low mass per unit area, protected from the radiation environment during transit to, and radiation resistant in operation in GEO. They will have to be producible at rates and in volumes consistent with an SPS deployment schedule. Finally, they will have to be integrable into arrays at a low cost per watt per unit area and consistent with transportation and system requirements. A single-crystal silicon solar cell array has been designed for use in the SPS (Figure 2) which incorporates a sawtooth cover glass to increase solar cell efficiency by about 1.5%.[3] Table 2 shows a mass breakdown of this solar cell array.

A typical single crystal silicon solar cell array panel would be about 1 m² in size and contain 252 silicon solar cells (18 connected in series and 14 in parallel). Annealing methods could be utilized to eliminate or reduce the degrading effects of accumulated radiation exposure and significantly extend the lifetime of the 50-μm solar cells.

Pulsed laser beams and scanning electron beams have been used[32] to anneal silicon solar cells. Substantial recovery of the performance of solar cells subjected to proton

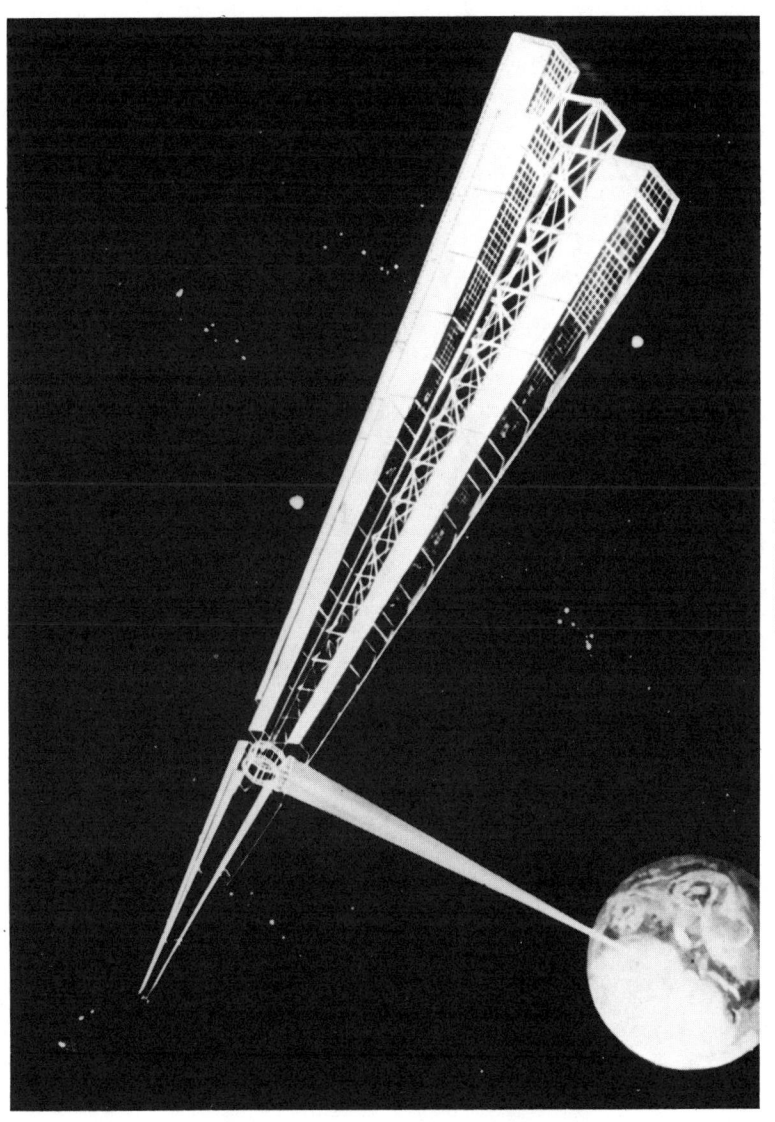

FIGURE 1(B).

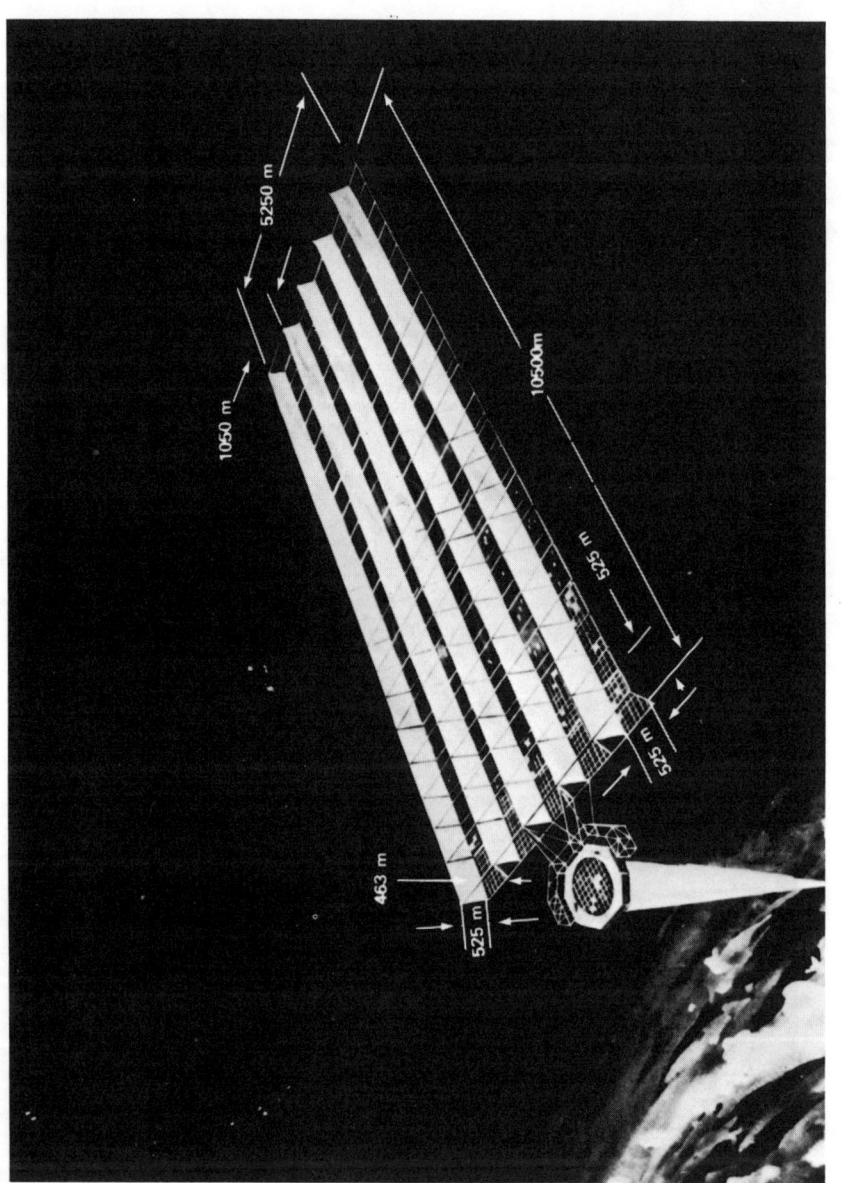

FIGURE 1(C).

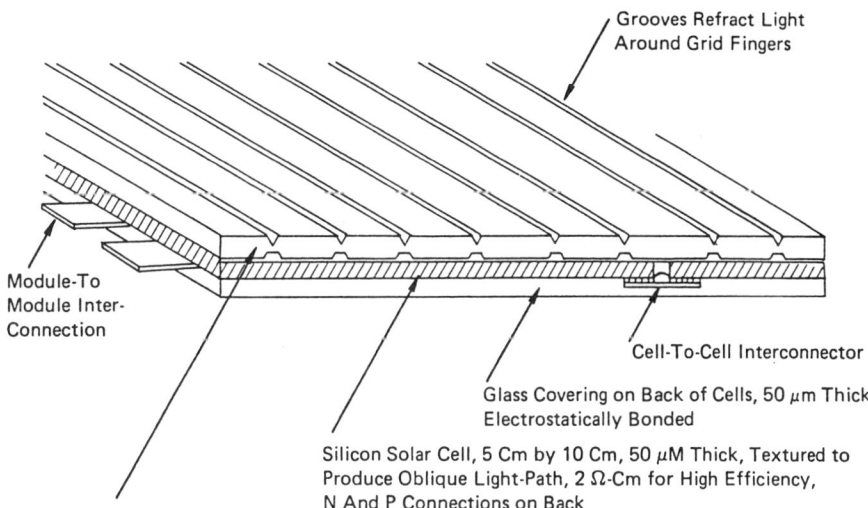

Grooves Refract Light
Around Grid Fingers

Module-To
Module Inter-
Connection

Cell-To-Cell Interconnector

Glass Covering on Back of Cells, 50 μm Thick,
Electrostatically Bonded

Silicon Solar Cell, 5 Cm by 10 Cm, 50 μM Thick, Textured to
Produce Oblique Light-Path, 2 Ω-Cm for High Efficiency,
N And P Connections on Back

Cell Cover of 75 μM Borosilicate Glass, Electrostatically
Bonded in High-Volume Equipment, Cesium Doped to
Give Ultraviolet Stability

Interconnectors: 12.5 μM Copper, With in-Plane Stress Relief, Welded to
Cell Contracts

FIGURE 2. Annealabel solar cell blanket structure (From Boeing Aerospace Co., Seattle, Wash., December 1977.)[3]

Table 2
SOLAR CELL ARRAYS: DESCRIPTION AND MASS

Component description	Mass (10^{-3}kg/m^2)
Si component description	
75-μm cover—fused silica	16.8
50-μm cell—Silicon	11.5
12.5-μm interconnects	1.1
50-μm substrate—fused silica	11.2
5% tolerances	2.0
Total	42.6
GaAs component description	
20-μm Al$_2$O$_3$	7.96
Interconnects	3.4
Grid contacts	0.03
0.03-0.05 μm GaAlAs ⎫	
1.5-μm p type GaAs ⎬ 5 μm	2.66
4.6-μm n-type GaAs ⎭	
0.5-1 μm ohm contacts	4.0
13-μm FEP	2.7
25-μm Kapton	3.6
6-12 μm polymer thin coating	0.9
Total	25.25

Courtesy of Boeing Aerospace Co., Seattle, Wash., December 1977,
and Rockwell International, Los Angeles, April 1978.

and electron irradiation has been demonstrated. It also has been shown possible to anneal radiation damage of silicon solar cells through integral glass covers. The annealing results suggest that with process optimization, very effective and possibly complete performance recovery will be possible. Using pulse energy heating to repair radiation damage will make it possible to anneal large surface areas very rapidly, i.e., at a projected rate of 5 m²/sec (150 km²/year). Adjustment of pulse conditions and scanning rates will ensure that the heating is limited to the solar cell and not transferred to the support structure. To maintain peak performance, the solar cell arrays may require annealing at 3- to 5-year intervals. The annealing process could also be used to repair those solar cells that may be installed in low Earth orbit (LEO)—i.e., to provide power for an electric propulsion system to transport the partially assembled SPS from LEO to GEO—after they have been exposed to the Van Allen radiation belts.

The solar cell panels are assembled into an installable 20 X 160 m package forming one of the bays of the SPS solar collector.[3] Loads within each bay are carried by a supporting grid and are applied through catenary supports attached to structural supports at intervals of 20 m. The load at each attachment point is designed to result in a frequency of about 12 cycles per hour, which is higher than the first few SPS support structure frequencies. Power distribution is accomplished with power sectors, each of which is switchable and isolatable from the main power bus to facilitate annealing or other maintenance functions. A rotary joint attaches the microwave transmitting antenna to the main structure. Power is transferred across the rotary joint by a slip ring-brush assembly with a diameter of 16 m. Mechanical rotation and pointing is accomplished by a mechanical turntable with a diameter of 150 m. To isolate the transmitting antenna from turntable vibrations, the antenna is suspended in a yoke by a soft mechanical joint. The antenna is mechanically aimed by thrusters distributed throughout the structure. A microwave interferometer of a type already demonstrated on the ATS-F spacecraft could be used to accurately sense the position of the transmitting antenna.

Table 3 shows the mass of a silicon solar array configuration of the SPS. The solar cell array design utilizes 50-μm silicon solar cells (6.5 × 7.5 cm) which are electrostatically bonded (after electrical interconnections are made) between a 75-μm cover glass and a substratum of 50-μm borosilicate glass.

2. Gallium Arsenide Solar Cells

Single-crystal heterojunction gallium arsenide (GaAs) solar cells have been demonstrated to achieve conversion efficiencies of 17% and a 20% efficiency is projected. Gallium arsenide solar cells have a substantial advantage because at elevated temperatures their efficiency does not degrade as fast as that of the lower band gap semiconductors; therefore, they can be used in systems utilizing concentrated solar energy. In addition, gallium arsenide solar cells are more resistant than silicon cells to radiation damage, thus promising a longer life, as well as higher performance in the space environment.

The commercial viability of thin-film gallium arsenide solar cells has not yet been demonstrated on a scale that even remotely approaches the SPS solar cell array requirements. Although good results have been obtained in the laboratory, pilot plant production based on promising fabrication processes will have to be demonstrated.[11]

The potential advantages of gallium arsenide solar cells for SPS applications have been recognized and developed for the SPS configurations shown in Figures 1B and 1C.[35] The solar cell consists of a 5-μm-thick layer of gallium arsenide deposited with a potentially low cost metal oxide vapor process onto a 20-μm-thick layer of synthetic sapphire, which becomes the cover glass for the inverted solar cell (Table 2). To reduce

Table 3
5-GW SPS SUBSYSTEM MASS COMPARISON FOR SI AND
GAAS SOLAR ARRAYS MASS-10⁶ kg

	GaAs (CR = 2)[a]	Si (CR = 1)[a]
Solar array	13.798	27.258
Primary structure	4.172	3.388
Secondary structure	0.581	0.436
Solar blankets	6.696	22.051
Concentrators	0.955	
Power distribution and conditioning	1.144	1.134
Information management and control	0.050	0.050
Attitude control and stationkeeping	0.200	0.200
Antenna	13.382	13.382
Primary structure	0.250	0.250
Secondary structure	0.786	0.786
Transmitter subarrays	7.178	7.178
Power distribution and conditioning	2.189	2.189
Thermal control	2.222	2.222
Information management and control	0.630	0.630
Attitude control	0.128	0.128
Array-antenna interfaces	0.147	0.147
Primary structure	0.094	0.094
Secondary structure	0.003	0.003
Mechanisms	0.033	0.033
Power distribution	0.017	0.017
Subtotal	27.327	40.787
Contingency (25%)	6.832	10.197
Total	34.159	50.984

[a]CR = Concentration ratio of solar radiation.

Courtesy of Boeing Aerospace Co., Seattle, Wash., December 1977, and Rockwell International, Los Angeles April 1978.

the required solar cell area, thin-film Kapton solar reflectors can be used at a concentration ratio of 2. An extension of this technology to multiple-band-gap solar cells with efficiencies of at least 25% can be projected. In addition gallium arsenide solar cells are more resistant to radiation damage and can be annealed at lower temperatures than silicon solar cells.

The mass of the solar arrays (Table 3) is the dominant component for the photovoltaic SPS designs. The scale of commitment of capital, material, and labor to construct large-scale manufacturing facilities to produce the solar array areas for the SPS will be substantial (Table 4). A comparison with areas required for a terrestrial photovoltaic system is shown in Table 1.

B. Thermal-Electric Conversion

Thermal-electric conversion is being considered as an alternative method of producing power in space. In this approach mirrors focus solar radiation into a cavity

Table 4
SOLAR ARRAY PRODUCTION REQUIREMENTS

Material	Concentration ratio	Total array area (km²)	Production rate[a] (m²/hr) 1 SPS/year	3 SPS/year	6 SPS/year
Silicon[b]	1	110.2	12,600	37,800	75,600
Gallium arsenide[c]	2	61.2	6,990	20,970	41,940

[a]100% yield, 24 hours/day, 365 days/ year.
[b]10-GW at utility interface: one Si SPS (CR = 1) with 110.2 km² of array.
[c]Two 5-GW GaAs SPS's (CR = 2) each with 30.6 km² of array.

Courtesy of Arthur D. Little, Inc., Cambridge, Mass., March 31, 1978.

where the heat is absorbed by a circulating fluid and transferred to heat engines coupled to electric generators. The electricity is then supplied to the power transmission system.[3]

Substantial terrestrial experience with applicable thermodynamic cycles indicates that thermal electric conversion in space could be accomplished by using heat engines operating on the Brayton or Rankine cycle. A Brayton-cycle heat engine with an inert gas as the working fluid does not require boilers and condensers, but for efficient operation it must operate at high temperatures, preferably higher than 1000°C. This will require the development of high-temperature materials, such as refractory alloys or ceramics, for turbine components and an optical system with a high concentration ratio, on the order of 1500, located perpendicular to the orbit plane. The efficiency of the space heat engine depends on the heat-rejection area ratio (radiation area to absorber area) and the temperature ratio of the thermodynamic cycle.

The mirrors focus solar radiation into a cavity absorber, where the heat is transferred to heat exchangers forming the walls of the cavity. In the Brayton-cycle heat engine helium is circulated through the heat exchangers and then expanded through the turbine powering both the compressor and the generator. A gas-to-liquid heat exchanger rejects the waste heat to a liquid-metal cooling fluid (NaK), which is pumped through space radiators. Heat pipes isolate the cooling loop to prevent freezing of the liquid-metal cooling fluid when temperatures drop during eclipses. Meteoroid bumpers protect the cooling loop from damage by meteoroid impacts.

A Rankine-cycle heat engine with potassium as the working fluid and operating at 1000°C is an alternative approach to the Brayton cycle and appears to be the more advantageous approach.

Although solar thermal-electric conversion is based on known and demonstrated thermodynamic principles, complex systems capable of large power outputs with few devices are required. By contrast, photovoltaic conversion relies on a large number of simple devices requiring advanced mass production techniques. Figure 3 compares the performance and Figure 4 the mass of alternative solar energy conversion technologies for the SPS. Gallium arsenide solar cells and an advanced Brayton thermodynamic cycle exhibit comparable performance and mass. At present, photovoltaic conversion for the SPS is favored but thermal engines deserve continuing consideration.

IV. TECHNOLOGY OPTIONS FOR POWER TRANSMISSION TO EARTH

To transmit the power generated in the SPS to Earth, there are two optional transmitting mediums: (1) a microwave beam or (2) a laser beam.

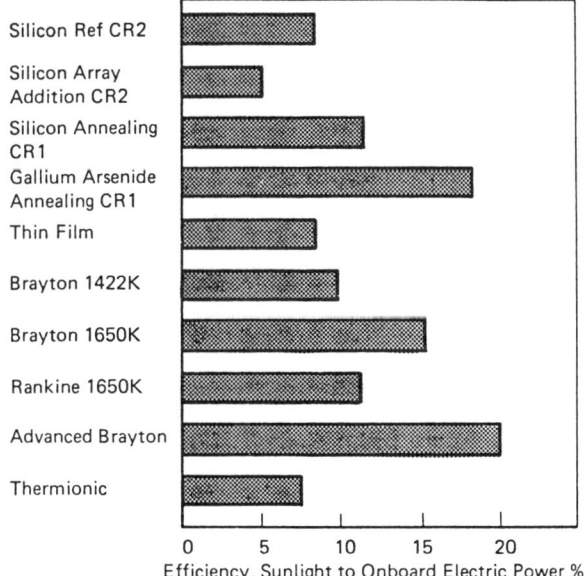

FIGURE 3. Performance comparison of solar energy conversion candidated for SPS. (From NASA, Lyndon B. Johnson Space Center, Houston 1977.)[3]

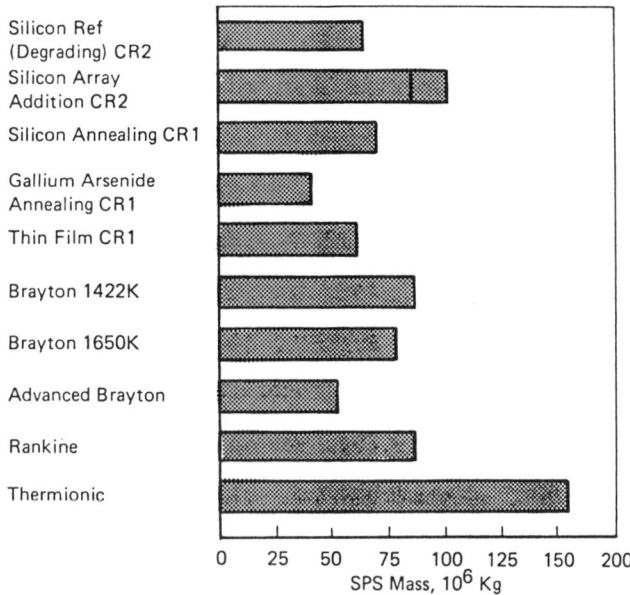

FIGURE 4. 10-GW SPS mass comparison of solar energy conversion candidated. From NASA, Lyndon B. Johnson Space Cener, Houston 1977.)[33]

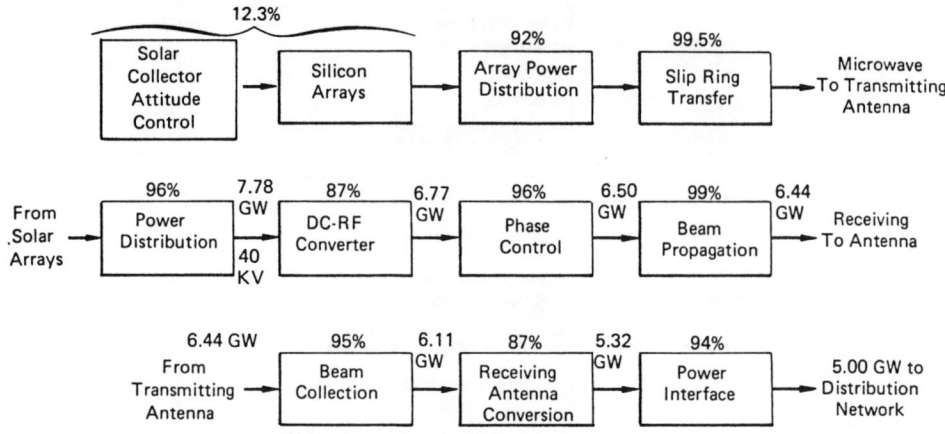

FIGURE 5. Microwave power transmission system: functional block diagram and efficiency chain. (From ECON, Inc., Princeton, N.J. March 1977.)[12]

A. Microwave Power Transmission System

Figure 5 shows the functional blocks of the microwave power transmission system, designed to transmit the electrical power generated by the solar energy conversion system to a receiving antenna on Earth, and the associated efficiency goals.[12] Free-space transmission of power by microwaves is not a new technology.[5] In recent years it has advanced rapidly, and system efficiences of 55%, including the interconversion between DC power and microwave power at both terminals of the system, are being obtained. The application of new technology is projected to raise this efficiency to almost 70%.

1. Microwave Power Generation

The devices that have been considered for converting DC voltage to RF power at microwave frequencies in the SPS are cross-field amplifiers (amplitrons) and linear beam devices (klystrons). The amplitron uses a cold platinum metal cathode operating on the principle of secondary emission to achieve a nearly infinite cathode life. With an output of 5 kW it could operate at an efficiency of 90%. The DC voltage required for the amplitron is 20 kV. The klystron could operate at an efficiency of 85% with an output of 70 kW but will require a more complex cooling system. Microwave solid state power transistors are also under consideration as a potential option for microwave power generation.

Considerations of mass, costs, and efficiency at specific frequencies have led to the selection of a frequency within the industrial microwave band of 2.40 to 2.50 GHz for the reference configuration. Alternative frequencies merit further study.

2. Microwave Beam Transmission

The transmitting antenna is designed as a circular, planar, active phased array having a diameter of about 1 km. Microwave power can be transferred at high efficiency when the transmitting antenna is illuminated with an amplitude distribution that is of the form $(1 - r^2)^n$ and when the phase front of the beam is carefully controlled at the launch point to minimize scattering losses.[5,18]

Space is an ideal medium for the transmission of microwaves: a transmission efficiency of 99.6% is projected after the beam has been launched at the transmitting antenna and before it passes through the upper atmosphere. To achieve the desired high efficiency for the transmission system while minimizing the cost, the geometric relationships between the transmitting and receiving antenna[19] indicate that the transmitting antenna should be about 1 km in diameter, while the receiving antenna should be about 10 km in diameter. The power density at the receiving antenna will be a maximum at the middle and will decrease with distance from the center of the receiver. The exact size of the receiving antenna will be determined by the radius at which the collection and rectification of the power becomes marginally economical. To reduce the dimensions of the transmitting antenna, the illumination taper from the center to the edge can be reduced at the expense of losing more power in the side lobes.

The transmitting antenna is divided into a large number of subarrays. A closed-loop, retrodirective-array phase-front control is used with these subarrays to achieve the desired high efficiency, pointing accuracy, and safety essential for the microwave beam operation.[19] In the retrodirective-array design, a reference beam is launched from the center of the receiving antenna and is received at a phase comparator at the center of each subarray and also at the reference subarray in the transmitting antenna center. The central subarray transmits the reference signal to the subarrays over calibrated coaxial cables (or by other means) so that its phase when it arrives at the phase comparator in each subarray is some integral multiple of 2π radians. The difference in phase between the signals, which, for example, may result from the displacement of a subarray from the nominal reference phase because of thermal distortion of the supporting structure, constitutes an error signal, which is used to correct the phase of the transmitted beam at the displaced subarray. Similar corrections in other subarrays result in the proper launching of the complete beam from the transmitting antenna array.

3. Microwave Power Reception and Rectification

The receiving antenna is designed to intercept, collect, and rectify the microwave beam into a DC output with high efficiency.[19,4] The DC output can be designed to either interface with high-voltage DC transmission networks or be converted into 60-Hz alternating current. The receiving antenna consists of an array of elements that absorb and rectify the incident microwave beam. Each element consists of a half-wave dipole, an integral low-pass filter, diode rectifier, and bypass capacitor. The dipoles are DC insulated from the ground plane and appear as RF absorbers to the incoming microwaves.

Because each dipole has its own rectifier, the receiving array has the directivity of a single dipole. Thus the collection efficiency of the array is relatively insensitive to substantial changes in the direction of the incoming beam. Furthermore, the efficiency is independent of potentially substantial spatial variations in phase and power density of the incoming beam, which could be caused by nonuniform atmospheric conditions.

The half-wave dipoles are spaced about 0.6 wavelength apart and are arranged in a triangular lattice at a distance from the ground plane of about one fifth of the wavelength. This distance may be adjusted within limits so that the receiving antenna provides a match between the specific DC load impedance and the incoming microwave beam. This match can approach 100%: reflection losses of less than 1% have been experimentally achieved. The low-pass filter is designed to minimize losses at the fundamental frequency and to reject effectively harmonics that are generated in the rectification process. The harmonics must be trapped and reflected in the proper phase to result in maximum RF-to-DC conversion. An RF-bypass capacitor acts as a smoothing filter to remove fundamental and harmonic ripple from the DC output: its position in relationship to the diode is such that a resonant circuit is created at the frequency of the incoming microwave power. Measured with specially developed equipment, the efficiency of the entire rectifier element has reached 90%. Field tests have demonstrated that an average rectification efficiency of 84% is achieved at the receiving antenna.

The amount of microwave power received in local regions of the receiving antenna can be matched to the power-handling capability of the microwave rectifiers. The rectifiers, which are Schottky barrier diodes made from gallium arsenide material, have a power-handling capability several times that required in the SPS application. Any heat resulting from inefficient rectification in the diode and its circuit can be convected by the receiving antenna to ambient air, producing atmospheric heating that will be only twice that of suburban areas, because only 15% of the incoming microwave radiation would be lost as waste heat. The low thermal pollution entailed in this process of rectifying incoming microwave power cannot be equaled by any known thermodynamic conversion process. The receiving antenna can be designed to be 80% transparent so that the surface underneath could be put to other uses. Receiving antennas could be located on land or offshore. A typical receiving antenna site will be 10×12 km. At least 100 potential sites have been identified in the U.S. Potential sites are more numerous west of the Mississippi: in the crowded Northeast, multiple use or off-shore sites may be needed.[6]

Albedo control at the receiving antenna site could compensate for solar radiation intercepted by the SPS and which would not have reached the Earth to maintain the global heat balance.

B. Laser Power Transmission

Microwave power transmission is the present choice, based on considerations of technical feasibility, fail-safe design, and low flux levels, but laser power transmission is an interesting alternative because of considerable advances in laser technology over the past 10 years and the possibility of delivering power in smaller increments to individual small receiving sites on Earth. The feasibility of high-power lasers for power transmission to Earth and the environmental, social, and economic implications are being evaluated.[2] Concentrated and dispersed beams generated by continuous-wave electric-discharge lasers—candidate lasants are carbon dioxide, carbon monoxide, mercury chloride, and mercury bromide—could be developed. (See Table 5.) Solar energy conversion, compatible with laser power transmission, may be carried out in GEO, or in lower orbits, particularly in a sun-synchronous orbit. If the latter orbit is to be used, a reflector will be required in GEO to reflect the laser power to a desired receiving site on Earth. The laser could be powered by photovoltaic or solar thermal conversion, or the solar energy could be used directly for laser pumping. Solar concentrators for laser pumping could be designed to focus only the portion of the solar spectrum that would be absorbed by the lasing species. For this purpose, coatings could be applied on plastic films, allowing the unused portion of the solar spectrum to pass through the solar collector.

The laser beam focus could be controlled with a pilot beam, phase locked to an adaptive laser beam projector to protect against beam wandering. The laser beam wavelength could be chosen so that nominal transmission losses would result under foggy or light cloud covers. Even if the receiving site were under heavy cloud cover, it could be interconnected with other sites to partially overcome weather-caused interruptions. Ninety percent of the continental U.S. has weather conditions that would permit efficient laser power transmission about 50% of the time.[2]

The following factors will affect the propagation of high-power laser beams: (1) linear absorption and scattering by atmospheric constituents; (2) atmospheric turbulence, induced random wander, spreading and beam distortion; (3) attenuation of the beam resulting from breakdown of the atmospheric gases; and (4) thermal blooming as a

Table 5
PARAMETERS OF STATE-OF-THE-ART AND CONCEPTUAL HIGH-POWER LASERS

Characteristics	State-of-the-art: Pulsed CO_2 EDL	Preliminary design: CW CO_2 EDL	Conceptual CW CO_2 EDL	Conceptual Projected HPL
Spectrum/range where band is located	10.6 μm	9.3 μm	10.6 μm	0.35—10.6 μm
Efficiency	0.23	0.235	0.20—0.30	0.83
Power output	175 kW	10 MW	75 MW	100 MW
Beam angle	1.2 × diffraction limited (without optics)	Same	Same	Same
Lasant operating temperature	300°K	350°K	300°K	500°K
Supporting equipment	Solar collector, DC resonant charging line pulser, lasant refurbishing equipment	Similar	Same	May be similar
Reliability	Must be very high[a]	Must be very high	Must be very high	Must be very high
Weight	203 kg/kW (average)	3.56 kg/kW	0.1 kg/kW	2—3.5 kg/kW
Related efficiencies				
Atmospheric transmission	0.80—0.90	0.80—0.90	0.88	0.95
Photovoltaic power collection system	0.151	0.151	0.151	0.151
Power receiver/converter	0.40—0.50	—	—	0.75

[a] Current units operate reliably for only a few minutes.

Courtesy of PRC Energy Analysis Co., McClean, Va., September 1978.

result of atmospheric absorption. The severity of these effects on laser power transmission will depend on atmospheric conditions at the receiving site, the wavelength of the radiation, the intensity, time characteristic of the beam, and the altitude of the receiving site.

Although laser power transmission is in an early stage of development, successful development of emerging technologies could make it a promising alternative to microwave transmission from space to Earth. Although high-power lasers could be used for military defense and pinpoint offense, compared to the destructive potential of nuclear weapons they are unlikely to be effective weapons of mass destruction. The potential for misuse of laser power transmission and hazards through its use will have to be investigated inasmuch as lasers may be perceived as either dangerous or, under certain conditions, provocative.

V. SPACE TRANSPORTATION SYSTEM

To be commercially competitive, the SPS will require a space transportation system capable of placing large and massive payloads into synchronous orbit at low cost. The cost of transportation will have a significant impact on the economic feasibility of the SPS. The space transportation system that will be available during the early phases of SPS development for technology verification and component functional demonstration will be the space shuttle, now well along in development and already demonstrated in flight. Compared to the previously used expendable launch vehicles, it will not only significantly reduce the cost of launching payloads, but will also be a major step toward the development of space freighters of greatly increased payload capability—and substantially lower costs.

The space freighter, which may be either a ballistic or winged reusable launch vehicle (HLLV), represents an advanced space transportation system with a planned capability to place payloads ranging from 200 to 500 metric tons into LEO. The space freighter will be recoverable and repeatedly reusable. The fuel for the lower stage will be liquid oxygen and a hydrocarbon: liquid oxygen and liquid hydrogen will be used for the upper stage. Both offshore and onshore launch facilities could be developed for the space freighter. Frequent launches (e.g., 10 launches per day) will necessitate maintenance and overhaul procedures similar to those employed in commercial airline operations.

The space shuttle will be adequate to meet the SPS development requirements over at least a 10-year period. With space freighter development started in the late 1980s, freighters would be available to launch the first commercial SPS after 2000.

Personnel and cargo will be transported from LEO to GEO with vehicles specifically designed for this purpose. The material required for the SPS construction and assembly will be transported by a cargo orbital transfer vehicle, which could be powered by ion thrusters of high specific impulse. Although the transit time to GEO would be measured in months, ion thrusters would minimize the amount of propellant to be transported to LEO.

Transportation costs of ballistic or winged-launch vehicles to LEO will be about $20/kg. including amortization of the vehicle fleet investment, total operations manpower, and propellant costs.[3] The total cost per flight will be about $8 million, with vehicle production and spares accounting for 40%, manpower for 35%, and propellants for 25%.

An efficient launch vehicle will consume about 23 kg of propellants to deliver 1 kg of payload to LEO. About 2 kg of propellant must be delivered to LEO for each kilogram delivered to GEO. Thus, to deliver two 5-GW SPS with a mass of 10^8 kg to GEO would require about 4.6×10^9 kg of propellants, divided as follows:[45]

1. 3.5×10^9 kg of liquid oxygen
2. 1×10^9 kg of hydrocarbons
3. 0.15×10^9 kg of liquid hydrogen

Assuming that 0.25 kg of oil is required to liquify 1 kg of oxygen, 5.25 kg of oil to liquify 1 kg of hydrogen, and 1 kg of oil for 1 kg of hydrocarbon, the oil equivalent to produce these propellants and to launch an SPS is about 2.7×10^9 kg or 19 million barrels. Thus, the launch of two SPS per year would require about 50,000 barrels per day or about 0.25% of current U.S. consumption.[45] In contrast, the SPS could produce power at the rate of 10^7 kW. At a 90% plant factor, this would be equivalent to a saving of about 400,000 barrels per day for 30 years or more.[45] This oil or its equivalent would have to be used to produce electricity throughout the operational life of the SPS if electricity were to be provided in a conventional power plant.

To construct four 10-GW SPSs per year would require about 10 launches per day. This level of operation, although orders of magnitude larger than present space experience, approaches the takeoff mass of the 500 to 700 flights per day of about 150 tons each at a large international airport. Experience with commercial airlines indicates that to service and support the launch vehicles would require only about 1200 personnel, although about 4700 personnel are assumed for the launch vehicles.[45] The major difference between the personnel required for the checkout of a reusable launch vehicle and the Saturn rocket is attributed to the total experience and operational history, which increases with each flight of a reusable vehicle, whether it is a commercial aircraft or a space transportation vehicle.

VI. ORBITAL ASSEMBLY AND MAINTENANCE

A. Technical Approach

The reduction of gravity and of the influence of forces shaping the terrestrial environment presents a unique freedom for the design of Earth-orbiting structures and provides a new dimension for the design of the structure required for the SPS, its fabrication, its assembly, and its maintenance in GEO. In GEO, the function of the structure is to define the position of subsystems rather than support loads that under normal operating conditions are orders of magnitude less than those experienced by structures on the surface of the Earth. The structure will have to be designed to withstand loads imposed during assembly of discrete sections, which may be fabricated in orbit and then joined to form continuous structural elements. The structure will therefore have to be designed to withstand both tension and compression forces that may be imposed during assembly and during operation when attitude control is required to maintain the desired relationship of the solar collectors with respect to the sun and of the transmitting antenna with respect to the receiving antenna on Earth.

The immensity of the structure alone ensures that it would undergo large dimensional changes as a result of the significant variations in temperatures that will be imposed on it during its periodic eclipses. During such eclipses temperature excursions as large as 100°C could be imposed, leading to substantial temperature gradients, which, depending on the dimensions of the structure, would cause dimensional changes of about 25 m for a 10-km-long solar array assembly if an aluminum alloy is used. Sudden temperature gradients during these eclipses would also cause lateral structural distortions that, depending on the structural damping characteristics, could propagate through the structure and set up oscillations that might persist for extended periods of time. To counteract the undesirable effects of large and sudden temperature changes methods to reduce thermal distortion and thermally induced stresses—methods such as thermal coatings, thermal shielding, and structural features—need to be further investigated.

The basic structural element in a triangular beam of required depths which provides the desired structural efficiency and low material mass can be fabricated at low cost by automated machinery. Both aluminum alloys and graphite composites show promise for use as the structural materials. Graphite fiber composites have a very small coefficient of thermal expansion compared to the aluminum alloys, but the aluminum structure could be insulated to reduce undesirable thermal effects.

The contiguous structure of the SPS is of a size that does not yet exist on Earth or in space. Therefore unique construction methods will be required to erect the structures that are used to position and support the major components, such as the solar arrays to form the solar collectors and the microwave subarrays to form the transmitting antenna. The basic approaches to constructing the required large space structure are as follows:[14] (1) deployable systems using elements fabricated on Earth, (2) erectable systems using elements fabricated on Earth, and (3) erectable systems using elements fabricated in space.

Deployable systems are widely used for the present generation of satellites, where the structure and components are packaged into the launch vehicle and deployed ready for use when the desired orbit is reached.

Erectable systems could be produced in the form of large-area structures that unfold at the desired orbit for assembly and whose sections then can be joined either manually or automatically. The methods for folding such erectable systems limit the compactness that can be achieved; thus they represent a penalty for launch vehicles that are volume limited rather than mass limited. This drawback is overcome when erectable systems are fabricated in space, which removes any limitation on the size of the structure that can be fabricated and subsequently assembled. Thus the size of the individual structural member, payload density, and overall mass of the completed space-fabricated structural element no longer constrain the design of assembly methods that can be developed for a large structure to be fabricated and assembled in space.

B. Construction Operations

The construction operations entail fabrication, assembly, and integration of the large solar energy conversion system (with sizes of the order of 5.20 km $\times$ 10.4 km for delivering 5 GW of power on the ground) and its microwave power transmission system antenna of about 1 km in diameter. Studies indicate that a 5-GW satellite could be constructed and checked out in GEO in 6 months. Construction of the solar arrays provides for in-orbit fabrication of lightweight structural beams together with their assembly into the framework for supporting the other subsystems. The overall construction process follows a repetitive production cycle. Automated construction techniques and equipment are used to minimize costly crew requirements during structural buildup and progressive installation and checkout of reaction control thrusters, high-voltage main power buses and jumper buses, high-voltage DC switch gear, solar arrays, instrumentation, and controls. A similar construction sequence is followed for the microwave antenna. It includes progressive fabrication and assembly of primary and secondary structures followed by installation and checkout of DC power distribution equipment, RF wave guide subarrays and power amplifiers, and supporting controls. A large rotary joint and yoke is also assembled to complete the electrical and mechanical interface between the solar arrays and the antenna. When the solar arrays and antenna are fully mated, final test and checkout will be automatically performed on the major satellite systems (e.g., flight control, DC power distribution, and RF phase control systems). Subsequent satellite power buildup operations will be controlled from the ground.

Automated machines capable of producing triangular truss shapes of desired sizes and material thickness in a modular configuration[31] could be employed. An automated beam builder consists of roll forming units that are fed with coiled strip or composite materials and automatically impart the proper shape to the individual strip, weld and fasten the individual elements, control dimensions, and produce the complete structural members. The prototype machine can produce a one meter deep structural member in increments of 1.5m at a rate of 0.5 m/min for continuous production of structural beams in space.

Warehousing logistics and inventory control will be required to effectively manage the flow of material to the SPS construction facility, which will be designed to handle about 100,000 tons per year. The construction facility could be a large lightweight rectangular structure with dimensions of about 1.4 × 2.8 km. It would provide for launch-vehicle docking stations and 100-person crew cylindrical modules with dimensions of about 17-m diameter × 23 m long.[45] The GEO construction facility will be designed to assemble the solar energy conversion system and the microwave transmission antenna.

Construction costs, including transportation of the required construction crew of about 550 people and amortization of the bases, are projected to account for about 8% of the total SPS capital cost. The construction crew's primary activity would be monitoring, servicing, and repairing, with little need for extravehicular activities. The SPS hardware throughput in the construction facility is projected to be 15 tons/hr for a construction rate of two SPS per year.[3] The repetitive automated production process of space construction activities is projected to result in a productivity per crew member of 10 man-hours per ton of materials handled (the experience with terrestrial steel construction projects). To reduce the cost of space construction the production process will have to be equipment-intensive rather than labor-intensive. Thus the significant capital investments can be amortized over a number of SPSs.

The SPS will be maintained on a scheduled basis by a maintenance crew using remotely operated machinery available at a maintenance base. The maintenance base would be shared among several SPSs to ensure that the supply of generated power would be up to the desired level at all times. Assuming appropriate maintenance the operational life of the SPS could be of indefinitely long duration because of the benign nature of the space environment compared to the terrestrial environment.

VII. SPS/UTILITY POWER POOL INTERFACE

The large power output potential of the SPS will require careful design of the utility power pool interface to reduce the impact on the stability of a total utility system. Electrical power grids are designed to provide this stability of power supply to the user by incorporating redundant installations of reliable equipment. At present, it is technically and economically feasible to construct a utility system that will, over a period of 10 years, meet its demand on all days except one, i.e., a loss-of-load probability (LOLP) of 0.1 day/year. The installed capacity required to provide an LOLP of 0.1 day/year using terrestrial equipment to meet typical loads is usually 25% greater than the peak yearly demand in any one power pool. The LOLP of the SPS will be influenced by unavoidable power losses during eclipses, failure of terrestrial equipment, and failure of spaceborne equipment. Failure of the ground equipment can be treated in a conventional manner. The probability of catastrophic failure of the SPS, for example, because of the impact of a large meteorite or of an accident will be very small and similar to that of a catastrophic failure of a power plant on the ground. The microwave transmitting antenna or station-keeping equipment, however, could fail in ways different from conventional failures and such failures could be serious. Therefore SPS maintenance will be performed at predetermined intervals and scheduled so that it would not occur near the period of peak demand.

In addition to mechanical reliability, the reliability of the SPS will depend on generic system effects and small variations (2 to 4%) caused by atmospheric absorption at the receiving antenna. Although the eclipse periods, occurring during the present periods of minimum demand are predictable outages, they are not planned outages since they are not deferrable. Thus, since they may affect the total system operation, they have to be included when calculating the forced outage availability of the SPS.

The stability of the SPS will have a substantial effect on the stability of the power pool that it serves. Low-frequency fluctuations could cause the power level delivered by the SPS to the receiving antenna to vary: high-frequency fluctuations could cause line surges that might disturb the transient stability of other generators in the power pool. The magnitude of these fluctuations will have to be investigated to establish the required degree of surge protection which would be supplied by short-term power storage (of the order of min) acting as a buffer. These issues inherent in the SPS utility interface represent a significant influence on specific design approaches and selection of technology options.

VIII. SPS IMPACT CONSIDERATIONS

A. Economic Considerations

The economic justification for proceeding with a solar power satellite development program is based on a classical risk/decision analysis which acknowledges that it is not possible to know the cost of a technology that will not be fully developed for at least 10 years, and the SPS plan calls for it to be commercialized, i.e., produced, operated, and maintained, in not less than 20 years. Justification, of course, is equally difficult to provide for other energy technology projects such as ocean temperature energy conversion (OTEC), the breeder reactor, and fusion. This justification, therefore, requires an appreciation of the competitive cost of alternative energy sources for the generation of electrical power that would be available in the same period.

Any SPS development program should be time phased so that the "economic" purpose of each program segment will be to obtain information that will permit the decision makers to make a deliberate decision to continue the program or to terminate it, and thereby to control the overall risk. Cost-effectiveness analyses alone would be inappropriate, as they would require postulating scenarios of the future that could be extremely difficult, if not impossible.

The benefits and cost of a development program as large as the SPS are not likely to be uniformly distributed but are more likely to be concentrated in certain segments of society and in the economy of industrialized nations. Individuals, corporations, institutions, and even entire sectors of industry will react to the cost and the benefits of the development as they perceive them. As a result of these perceptions, political pressures are likely to have a pronounced effect on the SPS development program, its schedule, and its ultimate success.

In the various studies to date, the major emphasis has been given to establishing technical feasibility: only limited economic feasibility studies have been performed, primarily pertaining to system costs, development program costs, costs of terrestrial alternatives, and comparative economic analyses of space and terrestrial power systems.[12,40]

1. Cost Projections

The economic viability of the SPS was compared with that of other alternatives to provide a basis for future decisions about a major SPS development program. The comparison indicated that if technology goals can be met, an operational 5-GW SPS would cost about $2600/kW once full production has been achieved and benefits of early experience have been incorporated into the design.[33]

The SPS cost estimates are based on point design and represent forecasts of future technology development that are unlikely to be precise. Risk analyses have been carried out to overcome the drawbacks associated with deterministic estimates. These analyses are based on the probable distribution of costs according to the present state of knowledge of the technology assumed for the SPS. Cost models were developed to determine unit production and operation and maintenance costs as a function of input variables.[12] A convergence of cost projections of the SPS indicated that capital costs would be in the range of $1600 to $3500/kW leading to electricity costs, based on a 30-year lifetime and a 15% return on investments as low as 30 mills/kWh, a nominal 60 mills/kWh, and 'an upper bound of 120 mills/kWh.[33] These costs lie within the competitive range of the costs of future terrestrial power-generation methods.

2. Institutional Impacts

Events have shown that controversies can arise over the utilization of existing energy technologies, even when they operate within well established performance and impact limits. In approaching the development of the SPS, the public response to its technology can be outlined only after the benefits and impacts of its performance are better defined. Although it is difficult to assess the institutional impacts of a concept like the SPS, which has not yet been demonstrated, even on a small scale, several issues are beginning to be explored and evaluated. Some of these deal with the potential damage to an SPS installation (which represents the concentration of massive amounts of capital and generating capacity) through accidents. Legal and political questions relate to impacts on telecommunications, both national and international, as well as the use of space and the rights of its use based on existing space law. A basic consideration will be the ownership of the SPS, the responsibility of the owners in case of accidents from whatever causes, and the vulnerability of the SPS to actions of adversaries.[13]

The existing body of technical and legal decisions[15a] relating to the use of GEO will strongly influence the development of the SPS. Two existing institutions that will have an impact are the United Nations and its Committee on the Peaceful Uses of Outer Space, including its technical and legal subcommittees, and the International Telecommunication Union, which the United Nations recognizes as the special agency responsible for establishing radio regulations, including the allocation of the radio frequency spectrum, registration of frequency assignments to avoid harmful interference between the radio stations of different countries, coordination of efforts to improve the use of the radio spectrum, and other associated purposes.

The 1967 treaty on "principles governing the activities of states in the exploration and use of outer space, including the moon and other celestial bodies," recognizes the "use of outer space for peaceful purposes."[37] Article I of the treaty states that the use of outer space "shall be carried out for the benefit and in the interest of all countries . . . and shall be the province of all mankind," and "outer space . . . shall be free for exploration and use by all states without discrimination of any kind on the basis of equality and in accordance with international law. . . ." This article also treats "potentially harmful interference" in the use of outer space and requires appropriate international consultations. These and other provisions of the treaty cover the basic international legal principles that govern the use of outer space and, therefore are applicable to the SPS and its use of GEO.

The U.S. position (1978), which also is applicable to SPS, has been stated as "The United States holds that the space systems of any nation are national property and have the right to passage through and operations in space without interference. Purposeful interference with space systems shall be viewed as an infringement upon sovereign rights."[38]

The body of emerging international laws dealing with the uses of GEO is expected to develop in a way that permits the use of new technology to optimize orbit utilization by future satellite systems. The International Institute of Space Law has already addressed the SPS legal issues at its meeting in Lisbon in 1975. Although it is difficult to chart the course of development of international space law, assessments of future space law development[47] are already beginning to address the issues that are raised and the extent to which existing space law and other international agreements are adequate to meet the objectives of an SPS, as well as international institutions that may have to be created so that its potential benefits will be globally available.

B. Environmental Impacts

The social costs of environmental impacts of this alternative large-scale power-generation system, including the land used for launch sites, alternate land use, and the aesthetic effects of such use, have to be established so that the benefits of each specific system approach can be weighed against potential dangers to human health, destruction of valued natural resources, and the intangible effects that may influence the quality of life.

The following analysis of the potential impacts and benefits of an SPS system is patterned after the requirements expressed in the Power Plant Siting Act of 1971, involving land use impacts, water resources impacts, air quality impacts, solid waste impacts, radiation impacts, and noise impacts.

1. Land Use[28]

The receiving antennas could be located on a wide variety of terrain, ranging from desert to farm land, and even in off-shore locations. In the U.S. about 100 locations have been identified as potential sites for a typical receiving antenna[6] with dimensions of about 10 km east-west and 12 km north-south; the exact dimensions will vary depending on the latitude of the site. The microwave beam flux density at the edges of a site of this size would be 0.1 mW/cm², which would require a 17-dB taper on the microwave beam.

Assuming that 100 5-GW SPSs were to be put into operation, 100 receiving antenna sites would be required. This number of SPSs would require launch sites, each about 21 km². Table 6 compares the land use impacts of the SPS with other power-generation methods.

2. Water Resources[28]

On the assumption that the 100 receiving antennas would be constructed using conventional approaches, about 1.7×10^6t of concrete would be used in the foundations. The concrete would require 2.3×10^5t of water; the water would have to be available or brought to the site during the construction phase. Construction operation could damage the terrain, increase water runoff during storms, and decrease the water supply to the local ecosystem. Construction in deserts could lead to modifications of the water cycle. Thus the impacts on water resources will have to be evaluated for each specific site.

The total nonrecoverable water use at the SPS launch complexes over a life cycle of 30 years for 100 5-GW SPSs would be about 6t/MW-year. The water would be used to produce the hydrogen and oxygen for rocket propellants and for rocket exhaust cooling. Hydrologic studies would be required to ensure adequacy of water supply from local sources at each launch site. By comparison, a coal-fired power-generating plant would require about 500 to 9200 t/MW-year.

Table 6
LAND USE IMPACT COMPARISONS

Power generation method	Land use (m²/MWe-yr)
Coal-fired steam[a]	3600
Light-water reactor[a]	800
Solar thermal conversion	3600
Photovoltaic conversion	5400
SPS[b]	985
SPS[c]	675

[a] Includes fuel cycle
[b] Microwave intensity of 0.1 m W/cm² at receiving analysis site perimeter
[c] Microwave intensity of 1 m W/cm²

Courtesy of Jet Propulsion Lab, Pasadena, Calif., 1978.

Table 7
HEAT RELEASE COMPARISONS OF NATURAL AND MAN-MADE ENERGY SOURCES

Sources	Power (W/m²)	Impact
Volcano	10^5	Global
Thunderstorm	100	2 cm/hr rainfall
Lake evaporation	100	Increased downwind rainfall
Large industrial city	600	Climate change
Suburban community	4	No detectable climate change
SPS receiving antenna	7.5	No climate change projected

From NASA, Washington, D.C., 1977.

The industries involved in the manufacture of the SPS would use about 250×10^6t of solid material resources. The total water pollutants from manufacture would be about 0.2t/MW-year, including acids, bases, dissolved solids, suspended solids, organics, and a measure of the biological oxygen demand and chemical oxygen demand. The total is small compared to the effluent from conventional electrical power plants. For example, a coal-fired steam electrical power plant has total water pollutants of 6.7 to 630 t/MW-year. A light-water-reactor power plant of 35% cycle efficiency would discharge about 1.8 MW$_{th}$ to its cooling water for each megawatt of electricity generated, and the pollutants in the cooling water would be transferred to either the surrounding local water or to the atmosphere. No cooling water is required at the receiving antenna because of the low heat release (see Table 7).

3. Air Quality[28]
The generation of electrical power with the SPS would not—at least, directly—produce any air emissions or pollutants. But air pollutants would be produced in the mining, processing, fabrication, assembly, and construction of the SPS, the receiving antennas, the space transportation system, and the launch site complexes. Air pollutants would also be formed during the launch and boost of SPS payloads to LEO and during transfer from LEO to GEO.

The total environmental releases to the air would be small compared to fossil fuel

Table 8
AIR POLLUTANTS PRODUCED
DURING CONSTRUCTION OF SPSs[a]

Pollutant	Amount (mt/MWe-yr)
Particulates	0.251
Sulfur oxides	0.0093
Carbon monoxide	0.093
Hydrocarbons	0.0226
Nitrogen oxides	0.0274
Ammonia	0.0006
Hydrogen fluoride	0.00065
Calcium fluoride	0.00011
Hydrogen sulfide	0.0005
Aldehydes	0.00010
Total	0.405

[a] 48 SPS, 10-GW output, constructed between 2000 to 2055.

Courtesy of Jet Propulsion Lab, Pasadena, Calif., 1978.

electrical power plants and comparable to the nonradioactive air emissions resulting from light-water-reactor power plant construction. Large amounts of particulates would result from cement production and cement use (Table 8) SO_2 would also be released during cement production and the production of steel and copper. CO would be released during the propulsion phase of booster flight and in the production of coke for steel and of thermal control coatings based on the use of carbon black. Hydrocarbons would be released in the production of coke and carbon black, and in the combustion of rocket propellants during launch to LEO. Nitrogen oxides would be produced during cement production and ammonia during the coke production process. Other pollutants would be released during the production of materials and during combustion processes. But the air releases of all those pollutants would be insignificant compared to those of coal-fired steam plants, which range from 5.5 to 110 t/MW-year of operation. Therefore, the effects on public health of SPS air pollutants are projected to be minimal.

4. Solid Wastes[28]

No solid wastes would be produced during the generation of electric power by the SPS. Solid wastes would only be formed during the manufacture and construction of the SPS, the receiving antenna and launch sites, and the space transportation system. They would amount to about 0.1t/MW-year, primarily attributable to aluminum, steel, and silicon production. The amounts are negligible compared to the 890 to 2100 t/MW-year from a coal-fired steam electrical plant.

5. Noise Impacts[28]

The primary noise impacts would be at the launch complexes during the frequent launches of the launch vehicles while the SPS are being constructed. Noise will also be generated during launches of various vehicles to supply the SPSs with expendables and for their maintenance, but these launches would be less frequent. Noise impacts during the construction of the receiving antennas would be minor because the sites would be remote from populated areas.

The launch complex sites would most likely be located in the less populated southwestern region of the U.S. An international cooperative effort, however, could lead to the selection of launch complexes nearer the equator where there are large uninhabited land areas, or in offshore locations.

6. Microwave Beam Effects

Atmospheric Attenuation and Scattering[34]—The atmospheric transmissions efficiency of the microwave beam depends on meterological conditions. Absorption by the atmosphere occurs when gaseous molecules with permanent dipole moments couple the electric or magnetic components of the microwave field to their rotational energy levels. Most of the absorption due to excitation of collision-broadened lines occurs at the 22-GHz line of water vapor and 60-GHz line of oxygen. Below 10 GHz the attenuation resulting from molecular absorption is approximately 0.1 dB or less, depending on the microwave beam elevation angle.

Attenuation of the microwave beam by rain, cloud droplets, snow, and hail will depend on their size, shape, and statistical distribution and composition. Rain, wet snow, melting precipitation, and water-coated ice attenuation is low at frequencies below 3 GHz. The most severe condition is expected in rain clouds, where attenuation may reach 4% at 3 GHz. The attenuation produced by a 1-km path through wet hail could reach 13% at 3 GHz.

Forward scattering by rain and hail will increase the field intensity outside the main microwave beam. For example, a 5-GW SPS operating at 3 GHz would scatter 3 mW nearly isotropically if the storm cell height were 1 km. At a range of 10 km, the scattered microwave beam power density would be about 2×10^{-4} mW/cm². Therefore, scattering by rain or hail is not expected to increase sidelobe levels significantly or broaden the main microwave beam.

Ionospheric Propagation—Among the several possible interactions of the microwave beam with the ionosphere are (1) *Ambient refraction of the microwave beam by the ionosphere*—this effect leads to a negligible displacement; if horizontal gradients are present in the ionosphere, they could result in displacements (less than 100 meters) of the microwave beam.[34] (2) *Ionospheric electron density irregularities*—these self-induced or ambient irregularities will cause phase fluctuations (less than 10 degrees) across the wavefront of the reference beam propagated from the center of the receiving antenna to the transmitting antenna face; random phase variations will subside within a few hundred meters and within tens of seconds.[34]

Power beam dispersion due to ionospheric density fluctuations will increase the field intensity at the beam edges by up to 30%. At low power densities these fluctuations at the edge of the beam will not cause any significant power loss.[34]

Experiments at Platteville, Colo., and Arecibo, P.R., indicated that microwave fluxes can produce significant changes in the thermal energy of the plasma in the D(*) (60-90 km), E (*) (90-150 km), and F (150-340 km) regions. However, the direct effect of microwave beam transmission with densities of 20-30 mW/cm² is likely to be small, inasmuch as the absorption at the 3-GHz frequency remains negligible, even with an order of magnitude increase in electron temperature and density. However, power densities greater than 100 mW/cm² could produce large horizontal electron density gradients that could cause significant beam displacement and dispersion. Results of recent experiments at Arecibo cast doubt that thermal runaway effects on the ionosphere will occur at these low microwave beam densities.[10] But thermal self-focusing and plasma striations could occur: so the magnitudes of these effects will have to be determined analytically and experimentally.

Faraday rotation effects relating the total polarization twist of a linearly polarized wave to the total columnar electron content of the ionosphere under geomagnetically quiet conditions are projected to produce insignificant polarization losses. During geomagnetically disturbed periods—severe geomagnetic storms occur about three times a year—Faraday rotation and polarization loss is projected to be less than 1%.[34]

7. Effects on the Atmosphere

The complex processes that occur naturally in the lower and upper atmosphere could be disturbed by SPS launch operations, debris generated during assembly and manufacture,[24] microwave beam interactions, and receiving antenna functions.

The physics and chemistry of the upper atmosphere and the processes that induce changes in atmospheric conditions, either by natural causes or by the release of effluents from aircraft and pressurized containers, are under intense investigations and have figured prominently in controversies pertaining to the depletion of the ozone layer by SST operations and the release of fluorocarbons into the atmosphere. Effects of rocket effluents on the upper atmosphere have led to ionospheric depletion, as demonstrated by the launch of Skylab[30] and by recent experiments to produce "holes" in the ionosphere, which led to a rapid large-scale loss of the plasma.[46] Repeated launches of the space freighter required for SPS deployment could have significant effects on the ionosphere over an area extending several thousand kilometers where the launch trajectory intercepts the ionosphere. The spatial and temporal extent of modifications of the ionosphere will depend on the number and rate of molecules injected by the propulsion system and the trajectory of the launch vehicles.

The potentially harmful effects of space shuttle exhaust in the stratosphere are receiving considerable attention.[7] Injections of water vapor and NO_x (which are involved in the complex sequence of chemical reactions governing the abundance of ozone in the region from 20 to 35 km) are projected to result in a reduction of the mean abundance of ozone, although there is still uncertainty regarding the roles of each of these components. The actual effects of any given rate of injection of either of these two components are difficult to determine because of uncertainties regarding the vertical and horizontal movements in the stratosphere which govern the rate at which they are injected, distributed, and ultimately removed from it; the lack of experimental observations on space vehicle emissions; the composition of the stratosphere as a function of altitude, location over the surface of the globe; and the nature of the chemical and photochemical reactions that determine the abundance of chemical species involved in the ozone equilibrium. Because vertical mixing in the stratosphere is very slow (about 2 years at 20 km and 4 to 20 years at 50 km) and declines with increasing altitude, gases injected into the stratosphere will accumulate even at a low annual rate of injection and could yield a large equilibrium value at very high altitudes.[18] (The region from 50 to 100 km contains only 0.1% of the total mass of the atmosphere.)

Although the chemistry of water vapor in the upper stratosphere has been studied, there is uncertainty regarding the possible consequences of incremental additions of water vapor. Water vapor is photodissociated to form radicals and molecules which will react with ozone and molecular and atomic oxygen. Futhermore, changes in the water vapor content could influence the natural flux of NO_x to the level of the ozone layer. Consequently, the effects of the space transportation system on water vapor injection, particularly in the upper stratosphere, require further investigation.

Exhaust products released in the upper atmosphere might also affect the lower atmosphere, either through migration or coupling mechanisms that are not yet fully understood, for example, the formation of nuclei for the condensation of water vapor causing high-altitude clouds which could modify the radiation balance of the Earth. Most of these effects could be mitigated if it were possible to design the orbit insertion trajectory of a space freighter to reduce or eliminate exhaust emissions during passage through specific regions of the upper atmosphere.

Ion engine effluents—i.e., argon ions, if argon is used as a propellant for orbital transfer vehicles between LEO and GEO—could alter the composition and dynamics of the ionosphere because of the very long charge-exchange lifetimes in the plasmasphere, where the argon ions would be trapped for hundreds of hours.[8] Interactions of solar and cosmic radiation and micrometeoroid impacts with SPS construction materials could erode surfaces at a rate of several hundred kilograms per day, resulting in clouds of gas and particles between the Earth and the Moon. Such clouds could reduce the solar radiation received by the SPS and also interact with the upper atmosphere.

The possibility of microwave transmissions through the troposphere locally heating and changing the atmospheric circulation dynamics is projected to be minimal. Furthermore, the projected heat release of 7.5 W/m^2 at the receiving antenna woud be less than 20% of the heat release of a conventional power plant and only twice the heat release over a suburban community. Therefore, the intensity of atmospheric disturbance due to receiving antenna operations will be very small compared with other man-made installations. Microwave beam heating of the lower atmosphere through gas absorption will be negligible, while scattering by the particles, even in a heavily polluted atmosphere, is not expected to be significant.

8. Effects on Health and Ecology
a. Microwave Biological Effects

The designs of the transmitting and receiving antennas are strongly influenced by the choice of the power distribution within the microwave beam and determine the level of the microwave beam power density at the edges of the receiving antenna site. Acceptable guidelines for continuous low-level exposure to microwaves must be used.

At present, there is a considerable difference of scientific opinion on the appropriate exposure levels to microwave radiation. Exposure guidelines adopted by several nations differ sharply. In the U.S., microwave radiation protection guidelines, first proposed in 1953, were based on physiological considerations—i.e., continuous whole body exposure of a human subject resulting in a maximum equilibrium temperature rise of 1°C. This guideline was adopted as a standard by the Tri-Service Committee in 1957 and was accepted by U.S. government agencies and industry as tolerable on a long-term basis without risk of irreversible damage. Experimental results on animals exposed to microwave radiation indicated that irreversible tissue damage occurred at power densities of about 100 mW/cm^2. On the basis of the premise that thermal effects predominate at levels higher than 10 mW/cm^2 and that below this power density level nonthermal effects predominate, the American National Standards Institute recommended in 1966 that 10 mW/cm^2 be used as an acceptable standard.

The U.S.S.R. microwave exposure standards rest on the empirical findings by Soviet scientists that microwave exposure could affect the nervous systems of animals and humans. Their studies indicated that the central nervous system is particularly sensitive to microwave radiation, and that chronic occupational exposure of humans to very low-power microwave radiation leads to a variety of psychological and physiological effects.

In contrast to these findings, the U.S. researchers who attempted to duplicate Soviet results found no acute transient cumulative psychological or physiological changes that could be attributed solely to microwave radiation exposure. The U.S.S.R. has adopted

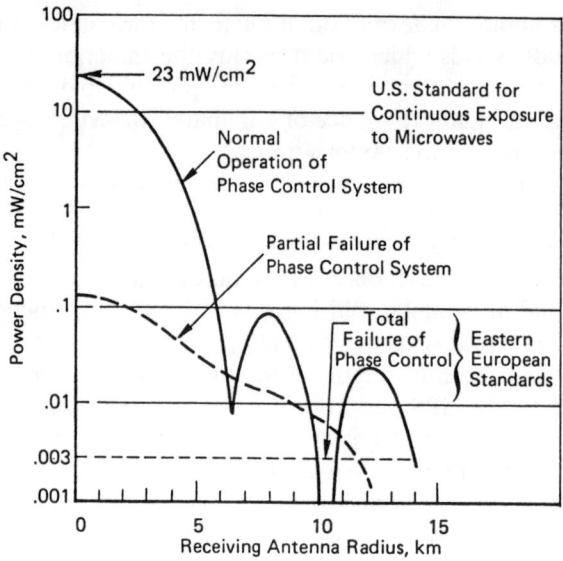

FIGURE 6. Effects of phase control on microwave power distribution at receiving antenna. (From NASA, Lyndon B. Johnson Space Center, Houston, 1977.)[33]

microwave radiation exposure standards that are 1000 times lower than the equivalent U.S. standards.

In view of this significant discrepancy, in 1968 the U.S. Office of Telecommunications Policy established the Electromagnetic Radiation Management Advisory Council, which obtained research funds for genetic, endocrinological, neurophysiological, and behavioral studies. The National Council for Protection and Measurements appointed committees of scientists, engineers, and physicians to study the effects of microwave radiation. In 1973 the Department of Health, Education and Welfare arranged a meeting in Warsaw under the auspices of the World Health Organization to resolve the discrepancies between the various microwave exposure standards.

In 1975 the International Union of Radio Science embarked on an expanded program to determine the biological effects of microwave radiation. Contemporary meetings of the International Microwave Power Institute, and the Institute of Electrical and Electronics Engineers are devoted to presentations on microwave radiation effects. Thus the subject of microwave biological effects has been and is receiving considerable attention from the scientific community, making it likely that this activity will lead to the adoption of international standards for continuous exposure to low-level microwaves.

In view of the current interest in low-level effects of microwave radiation, it is very likely that international standards will be adopted in time to be incorporated in the SPS design. The basic premise is that the SPS microwave transmission system must be designed so that the microwave power flux densities to which the public would be exposed outside the receiving antenna site will meet international standards.

The SPS microwave transmission system design incorporates the principle of retrodirective control of the microwave beam to make it impossible for the beam to be pointed to any location but that of a receiving antenna and instantaneous shutoff of power fed to the microwave generators, resulting in graceful system failure modes (Figure 6).

Until acceptable international standards have been agreed on, a microwave power flux density of 0.1 mW/cm² at the perimeter of the receiving antenna site has been assumed for the system design studies. The effects of either an increase or decrease in

Table 9

RANGES AND LIMITS OF POWER DENSITY FOR MICROWAVE EXPOSURE

Impacted biota	Guideline limit	Expected maximum
Public	≤ 1.0 mW/cm² (continuous)	≤ 1 mW/cm² (outside exclusion zone)
Terrestrial worker	≤ 1.0 mW/cm² (continuous)	~ 23 mW/cm² (inside exclusion zone) may increase to ~ 100 mW/cm² under rectenna fault conditions)
	No stated limit for intermittent exposure	
Space worker	≤ 1.0 mW/cm² (continuous) ≤ 10.0 mW/cm² (per 8–hr day)	≤ 3500 mW/cm² (near field on axis)
Ecology	None	~ 23 mW/cm² (inside exclusion zone) ≤ 1.0 mW/cm² (outside exclusion zone)

From U.S. Department of Energy, Washington, D.C., 1977.

the permissible microwave power flux density based on international standards can be evaluated on the basis of system design parameter and economic factors. Table 9 shows the developmental and evaluation guidelines.[41] A research plan and schedule to assess the health and ecological impacts of microwave power transmission have been prepared[42] with the objective of completing the first phase of the plan by 1980.

b. General Health and Safety Effects

Most of the public health and safety effects attributable to the construction and deployment of an SPS result from conventional processes such as the extraction, processing, and fabrication of materials; the construction of equipment required for mining, processing, and manufacture; and the transportation of the materials and equipment to meet SPS requirement.[20] The increase in these activities can be judged from the materials requirements of the SPS (Table 10).[26] For example, substantial incremental additions to current production rates of hydrogen, bauxite, and tungsten would be required.

Mercury use can be eliminated by the choice of other heat transfer material to cool klystrons. Nonconventional activities are associated with the launch and recovery of the space transportation system.

The health and safety effects of these terrestrial operations include the accidental release of hydrogen during transportation to the launch sites and launch and recovery accidents during failures of the space transportation systems. The occupational health impacts on terrestrial workers for both conventional and nonconventional activities will have to be mitigated by close adherence to occupational safety and health standards.

The workers making up the crew of the space construction bases in LEO and in GEO will be exposed to effects that are unique to the space environment and to the operations to be carried out by them in that environment (Figure 7). Although the ability of man to perform effectively in space for periods of up to 140 days has been demonstrated in the 1978 Salyut program, the physiological and psychological effects of prolonged exposure to the space environment remain to be established. For example, the biological effects of high-energy heavy-ion components of galactic cosmic radiation, if significant, would not be eliminated by shielding methods currently under consideration. Interactions of

Table 10
MATERIALS AVAILABILITY/DEMAND ANALYSIS FOR TWO 5-GW/YR SPS WITH SILICON ARRAYS

Material	Estimated annual SPS requirement (t)	Domestic resources (t)	Domestic production capacity (t)	Annual production shipments (approximately 1973—77) (t)	Annual requirement estimated as percent of product shipped
Aluminum	284,642		4,794,400	3,809,400	5.94
Argon	18,690		317,450	243,500	5.89
Arsenic	7		37,200	NA	NA
Bauxite	1,437,442	275×10^6 (approximately)		1,804,400	79.66
Cement	330,225		86,979,000	75,179,400	0.38
Ceramics	103			1×10^{12}	—
Coal		$3,600 \times 10^9$		542×10^6	—
Concrete	2,660,000				
Copper	13,299	193×10^6	2,645,000	1,394,000	0.50
Gallium	7	4,500	8	NA	NA
Gallium arsenide	14			NA	NA
Glass	39,196			2,000,000	1.96
Graphite	9,429			NA	NA
Graphite epoxy	12,572			NA	NA
Hydrogen	128,547		Est. 30,000	19,087	673.48
Mercury	168	31,000	1,650	797	21.08
Methane	651,599	$4,620 \times 10^6$		374.1×10^6	0.17
Molybdenum	4	16.8×10^6		51,351	—
Oxygen	2,728,506		22,675,000	10,955,804	12.03
Silicon, metallurgical	16,078		160,000 (approximately)	11,560	14.41
Silver	76	177,300	1,493	1,068	7.12
Steel, stainless	11,102			1,527,776	0.73
Steel, structural	2,984,000			43,934,100	6.79
Titanium	248	224×10^6	1,063,900	19,954	1.24
Tungsten	1,220	434,500	4,213	2,725	44.77

Courtesy of PRC Energy Analysis Co., McClean, Va., 1978.

CAUSE   EFFECT

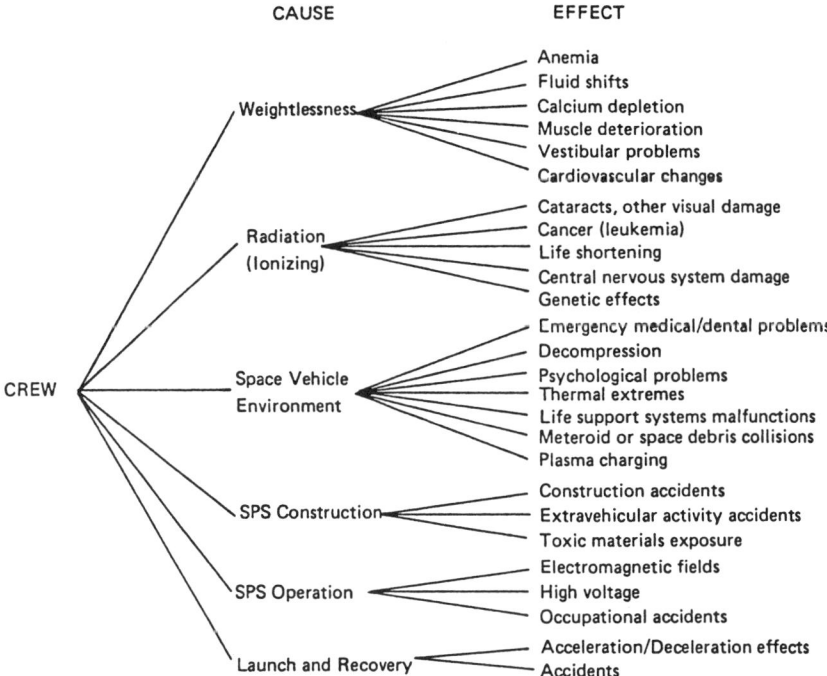

FIGURE 7. Potential health and safety impacts on space construction base crew. (From NASA, Lyndon B. Johnson Space Center, Houston, 1977.)[33]

the energetic heavy ions with shielding materials may produce secondary particles that may have deleterious biological effects. Protocols and procedures to ensure health maintenance and health surveillance of space workers will require additional data on space-related health problems and their mitigation by appropriate mission duration, crew rotation, and preplacement medical evaluations.[25] Selection of the desired life support systems and criteria to meet nutritional requirements are being recognized as worthy of increasing study.[36]

9. Radio Frequency Interference

Worldwide communications are based on internationally agreed on and assigned frequencies. Because the frequency bands spanning the most desirable operating frequency of the SPS are already in heavy use, the potential for radio frequency interference (RFI) of the SPS with existing communication systems is high. The microwave generators will have to be designed to filter out most spurious outputs.[18] RFI could occur during the shutdown of the microwave generators or result from fundamental microwave frequencies and their harmonics, random background energy, and other superfluous signals. Although RFI can be controlled by the selection of frequency, narrow-band operation, and use of filters, detailed and specific effects and impacts on radio astronomy, shipborne radar, and communication systems will have to be determined before the international acceptability of specific frequency allocations can be assured. The RFI effects and international agreements on frequency assignments are issues that will have to be faced at various stages during the SPS development and at meetings such as the World Administration Radio Conference, Geneva, 1979.

10. External Energy Subsidies

In addition to other economic comparisons, the external energy subsidies that are required to place an SPS in operation have to be considered, inasmuch as they

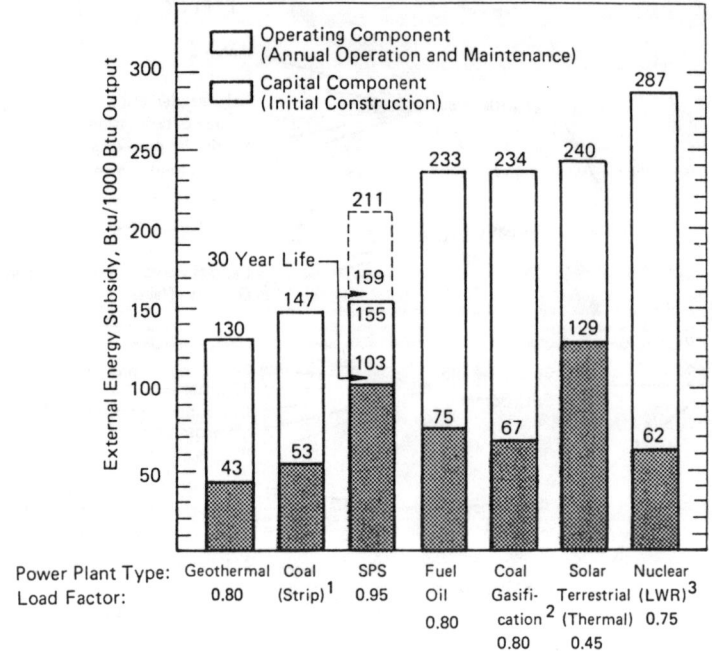

FIGURE 8. Comparison of external energy subsides for alternative power plants: (1) plant at mine mouth; (2) conversion and power plants both at mine mouth (Lurgi process); (3) includes modifications to original Development Sciences, Inc., report. Note that 20-year life is assumed unless otherwise noted. (From ERDA, 1976.)[15]

determine how long a power plant would have to operate before all of the energy required—i.e., to process the materials, fabricate and assemble components, and place the power plant in operation, including space transportation and the required propellants—would be paid back during normal operations. Figure 8 shows the external energy subsidies for a variety of present and future power plant.[15] The shaded area of the bar represents the subsidy required for initial construction (capital subsidy) and the unshaded portion of the bar indicates the subsidy required for annual operation and maintenance (operating subsidy).

On an energy subsidy basis, achievement of the system design goals would make the SPS very competitive with other systems. The results of the analyses of external energy subsidies based on even a limited 30-year lifetime of the SPS can be translated into a 3.25-year energy payback time for the SPS, including the operating subsidy. Energy impact assessments indicate that energy payback will be achieved in less than 1.6 years.[28] Evaluation of energy ratios on a "fuel-included" and "fuel-excluded" basis indicated that the SPS seems to be more efficient than coal or nuclear technology in utilizing fossil fuels to produce electricity.[21]

IX. SPS DEVELOPMENT PROGRAM

The projected scale of SPS development and operation, the financial and material requirements, the economic and social consequences, the international and political significance, and the magnitude of potential benefits on a national and international level place the SPS in the highest rank of socially sensitive technology programs. Because of the potential magnitude of the total development effort and the financial commitment that would be required, SPS rivals nuclear fission and fusion, satellite communications, and intercontinental aviation in significance; its development will require a similar scale of effort and time before SPS operations have any substantial impact on other energy resources. Therefore, it is essential to divide the SPS development program into well-defined phases so that limited objectives, in proportion to allocated funds, will have been reached at the conclusion of each phase. Decisions can then be made on the development objectives and procedures for the next and succeeding phases.

The stage is set to embark on a more intensive evaluation of the SPS option, including key terrestrial tests required to support system studies, to define the SPS development and operational phases, and to initiate supporting space experiments that will provide crucial information on which to base decisions concerning required technology and its impacts. The objectives of the near-term SPS development program are to (1) identify and assess issues that could constrain successful SPS development, and (2) seek ways to resolve these issues with a combination of analyses, system studies, and experiments on Earth and in space. The SPS development program can be divided into three overlapping phases, as follows:

Phase 1: *Concept Feasibility.* The objective of the present studies, which started in 1972, is to establish the overall feasibility of the SPS concept through system definition studies and environmental and socioeconomic assessments so that development program directions can be defined.[48]

Phase 2: *Technology Advancement.* Significant advancement of the technology for the SPS will require laboratory investigations, terrestrial testing, limited space experiments, and continuing in-depth evaluation of environmental effects, economic factors, and institutional arrangements to reduce program risks and uncertainties and to define future SPS development program directions.

Phase 3: *Demonstration Projects.* Demonstrations of the functions of critical elements and operational readiness of the SPS will require space projects to be carried out on an appropriate scale and with increasing space capabilities to provide information necessary for a decision to proceed with a full-scale demonstration program of the SPS. Demonstration is projected to be achievable by 2000.

Once the SPS technology has been demonstrated in an orbital flight test program, other countries may become interested in joining in the development of the operational SPS. International participation would permit the sharing of substantial development costs and ease obtaining international agreements, including frequency assignment and synchronous orbit positions, as well as provide assurance of the SPS's peaceful nature and its adherence to environmental standards. The spreading of the benefits of the SPS throughout the world will be important for its public acceptance, and therefore the SPS is receiving consideration by the United Nations.[37]

X. CONCLUSIONS

The potential of the SPS to meet future energy demands is being recognized,[22] and plans for its development are being studied. The results of extensive SPS system studies have confirmed that there are no known technical barriers to the design, deployment, or operation of the SPS. Economic studies have shown that projected capital and electric power-generation costs are within a competitive range of the costs of future terrestrial power-generation methods. Risk analyses have provided an economic justification for proceeding with the initial phases of an SPS development program. Environmental impacts have not emerged as a major constraint on SPS operation. There is growing confidence in the international technical community that the SPS is one of the most promising power generation options which could contribute to meeting global energy demands in the 21st century. Therefore, this option deserves serious consideration as humanity faces the challenges posed by the inevitable transition to renewable energy sources and becomes aware of the potential of space for improving the human condition.

REFERENCES

1. Evaluation of Solar Cells and Arrays for Potential Solar Power Satellite Applications, Final Report. NASA Contract No. NASA9-15294, Arthur D. Little, Inc., Cambridge, Mass., 1978.
2. **Bain, C. N.**, Potential of Laser for SPS Power Transmission, DOE Contract No. EG-77-C-01-4024. PRC Energy Analysis Co, McClean, Va., 1978.
3. Solar Power Satellite System Definition Study, Final Reports, NASA JSC Contract NAS9-15196, Boeing Aerospace Co., Seattle, Wash., 1977.
4. **Brown, W. C.**, Adapting microwave techniques to help solve future energy problems, *IEEE Trans. Microwave Theory Tech.,* 21(12) 753, 1973.
5. **Brown, W. C.**, The technology and application of free-space power transmission by microwave beam, *Proc. IEEE,* 62, 11, 1974.
6. **Chapman, P.**, Implications of Receiving Antenna Siting for SPS System Design, Final Report, NASA Contract No. NAS8-33002, Arthur D. Little, Inc., Cambridge, Mass., 1978.
7. **Ching, B. K.**, Space power systems—what environmental impact?, *Astronaut. Aeronaut.,* 15 (2), 60, 1977.
8. **Chiu, T. Y., Ching, B. K., and Luhmann, J. G.**, Global Scale Impact on the Ionosphere and Magnetosphere of Ion Engine Effluents from Solar Power Satellite Systems, Aerospace Corp., Space Science Laboratory, Los Angeles, Calif., 1978.
9. **Christol, C. Q.**, Satellite Power System (SPS) International Agreements, HCP/R-4024-08, U. S. Department of Energy, Washington, D.C., 1978.
10. **Coco, D. S. Duncan, L. M., and Showen, R. L.**, An experimental study of electron heating in the lower ionosphere, Presented at Int. Union Radio Sci. Natl. Radio Sci. Meet., Boulder, Colo., 1978.
11. **Dupuis, R. D., Dapkus, P. D., Yinphius, R. D., and Moudy, L. A.**, High efficiency Ga AIAs/GaAs heterostructure solar cells grown by metalorganic chemical vapor deposition, *Appl. Phys. Lett.,* 31 (3), 201, 1977.
12. Space-Based Solar Power Conversion and Delivery Systems Study, Final Reports, NASA MSFC, Contract NASS-31308, ECON, Inc., Princeton, N.J., 1977.
13. Political and Legal Implications of Developing and Operating a Satellite Power System, Final Report, JPL Contract 954652, ECON, Inc., Princeton, N.J., 1977.
14. **Engler, E. E. and Muench, W. K.**, Automated space fabrication of structural elements, in *The Industrialization of Space,* Vol. 36, (Part 1), Univelt, San Diego, 1978, 27.
15. ERDA, Task Group on Satellite Power Stations, ERDA-76148, Washington, D.C., 1976.
15a. **Gehrig, J. J.**, Geostationary orbit-technology and law, 27th Int. Astronaut. Congr., Anaheim, Calif., IAF-ISL-76-30, 1976.
16. **Glaser, P. E.**, Power from the sun: its future, *Science,* 162, 857, 1968.
17. **Glaser, P. E.**, Method and apparatus for converting solar radiation to electrical power, U.S. Patent 3.781.647, 1973.
18. **Glaser, P.E., Maynard, D. E., Mockovciak, J., Jr., and Ralph, E. L.**, Feasibility Study of a Satellite Solar Power Station, NASA CR-2357, NTIS N74-17784, 1974.
19. **Goubau, G.**, Microwave power transmission from an orbiting solar power station, *J. Microwave Power,* 5 (4), 233, 1970.
20. **Halverson, S. L., Rote, D. M., Rusch, C. M., Davis, K., and White, M.**, Preliminary assessment of the environmental impacts of the satellite power system (SPS), Int. Union Radio Sci., Natl. Radio Sci. Meet., Boulder, Colo. 1978.
21. **Herendeen, R.**, Two technologies near the energy limit, gasohal and solar power satellite power stations, presented at Symp. Inst. Gas Technol., Colorado Springs, Colo., 1978.

22. H.R. 10601, Bill to Create the Solar Power Satellite Research, Development and Program Act of 1978 (January 30, 1978), approved by the House of Representatives, 95th United States Congress, June 19, 1978.
23. Technical Bases for the World Administrative Radio Conf. 1979, Report for the WARE 1979, Int. Telecommunication Union, Int. Radio Consultative Committee, Geneva, 1979.
24. **Kessler, D. J. and Cour-Palais, B. G.,** Collision frequency of artificial satellites: the creation of a debris belt, *J. Geophys. Rev.,* 83 (A6), 2637, 1978.
25. **Kornberg, J. P., Chapman, P. K., and Glaser, P. E.,** Health maintenance and health surveillance considerations for an SPS space construction community, AAS 78-176, Am. Astronaut. Soc. Conf., Houston, 1978.
26. **Kotin, A. D.,** Satellite Power System Resource Requirements, DOE Contract No. EG-77-C-01-4024, PRC Energy Analysis Co., McLean, Va., 1978.
27. **Lindmayer, J., Wrigley, C., and Storti, G.,** Development of an Improved High Efficiency Thin Silicon Solar Cell, Quarterly Report, JPL Contract No. 954883, Solarex Corp., Rockville, Md., 1978.
28. **Livingston, F. R.,** Satellite Power System Environmental Impacts, Report 900-822, Rev. A, Jet Propulsion Lab, Pasadina, Calif., 1978.
29. **Meinel, A. B. and Meinel, M. P.,** *Applied Solar Energy,* Addison-Wesley, Reading, Mass., 1976, 108.
30. **Mendillo, M., Hawkins, G. S., and Klobuchar, J. A.,** An Ionospheric Total Electron Content Disturbance Associated with the Launch of NASA's Skylab, AFCRL-TR-74-0342, Air Force Cambridge Research Labs, Hanscom AFB, Mass., 1971.
31. **Miller, K. H.,** Solar Power Satellite Construction Concepts, in *The Industrialization of Space,* Vol. 36, part 1, Univelt, San Diego, Cal., 1978.
32. **Minnuci, A., Matthei, K. W., Kirkpatrick, A. R., and Oman, H.,** *In situ* annealing of space radiation damage, 13th IEEE Photovolt. Specialists Conf., Washington, D.C., 1978.
33. NASA, Lyndon B. Johnson Space Center, Solar Power Satellite Concept Evaluation, Activities Report, JSC-12973, 1977.
34. Microwave Power Transmission System Studies, Final Report, Vol. II, NASA CR-134886, Raytheon Co., Waltham, Mass., 1975.
35. Rockwell International Satellite Power System Concept Definition Study, Final Report, NASA MSFC, Contract No. NAS 8-32475, Rockwell International, Los Angeles, Calif., 1978.
36. **Spurlock, J. M. and Modell, M.,** Technology Requirements and Planning Criteria for Closed Life Support Systems for Manned Space Missions, Final Report, NASA Contract No. NASW-2981, Society of Automotive Engineers, Aut. England, 1978.
37. United Nations Committee on the Peaceful Uses of Outer Space, Solar Power Stations in Space, Background paper prepared by the Secretariat, United Nations, General Assembly, A/AC-105 CXIX CRP-1, 1976.
38. U.S. Position, Presidential Directive to Establish National Policies to Guide the Conduct of U.S. Activities In and Related to Space Programs, 1978.
39. Photovoltaic Program Summary, DOE ET-0019/1, Division of Solar Technology, U.S. Department of Energy, 1978.
40. U.S. Department of Energy, Satellite Power System (SPS) Concept Development and Evaluation Program Plan, DOE-ET-0034, NASA and U.S. Department of Energy, Washington, D.C., 1978.
41. U.S. Department of Energy, Interim Environmental Guidelines for Satellite Power System Concept Development and Evaluation, SPS Program Office, Office of Energy Research, Washington, D.C., 1978.
42. U.S. Environmental Protection Agency, Recommendations for a Program Plan to Assess the Health and Ecological Impacts of Microwave Power System from a Satellite Power System, Research Triangle Park, N.C., 1978.
43. **Vanke, V. A., Lopuhkin, V. M., and Savvin, V. L.,** Satellite solar power stations, *Sov. Phys. Usp.,* 20 (12), 1977; *Am. Inst. Phys.,* 0038-5670/77/2012-0989, p. 989, 1978.
44. Watt Engineering Ltd., On the Nature and Distribution of Solar Radiation, HCP/T2552-01, U.S. Department of Energy, Washington, D.C., 1978.
45. **Woodcock, G. R.,** Large-scale space operations for solar power satellites, AIAA/EEI/IEE Conf. New Options Energy Technol. San Francisco, Calif., 1977.
46. **Zin, J., Sutherland, C. D., and Ponpratz, M. B.,** Predicted Ionospheric Effects from Launches of Heavy Lift Rocket Vehicles for the construction of Solar Power Satellites, Report LA-UR78-1590, Los Alamos Scientific Lab, Los Alamos, N.M., 1978.
47. **Galloway, E.,** The Future of Space Law, 27th Int. Astronaut Congr., IAF-ISL-76-06, Aneheim, Colo., 1976.
48. U.S. Department of Energy, A Bibliography for the Satellite Power System (SPS) Concept Development and Evaluation Program, DOE/ER-0098, Solar Power Satellite Project Division, Office of Energy Research, Washington, D.C., April, 1981.

Chapter 4

SPACE MANUFACTURING OF NONTERRESTRIAL MATERIALS

Brian O'Leary and Gerard K. O'Neill

TABLE OF CONTENTS

I. INTRODUCTION

Five years have passed since the first publication on the practical use of nonterrestrial materials in space for the improvement of the human condition.[1] A number of factors combine to make such a concept apparently feasible within the next 15 to 20 years. Among them several are facts of nature independent of technology: the shallow gravity wells of the moon and asteroids; the presence (though unknown until 1969) of an abundance of glass, metals, and oxygen in the lunar soil; the availability in space of intense, full-time solar energy. Of technological factors, the most important has already begun operation: the space shuttle. Others have been planned to the level of basic engineering in a series of NASA-sponsored studies at the Ames Research Center in Moffett Field, Calif.[2-4] Other factors include the low cost transport of nonterrestrial materials to a stable high earth orbit by means of a solar-powered electric reaction engine (mass driver); the chemical processing of those materials into structures such as power stations[5] and, eventually, permanent human settlements as much as several kilometers across, each containing thousands to millions of inhabitants.

It is essential to this "High Frontier" concept that rapid exponential growth of industry and wealth can occur without adverse environmental consequences in the space environment, which has vastly greater resources of clean energy and of materials than does our Earth. It is also essential that no basic scientific breakthroughs appear necessary in reaching the new frontier.

More recently the possibility of constructing satellite power stations from nonterrestrial materials has been investigated as a first goal for space manufacturing.[5] The concept of supplying electric energy to the power grid here on Earth by radio transmission from a satellite located in permanent sunlight was proposed some 10 years ago by Dr. Peter Glaser.[6] Its attractions are that its energy source is continuous and inexhaustible, that it appears doable by engineering rather than requiring new scientific discovery, and that it appears far more acceptable environmentally in the long term than coal or nuclear power.

A task force at the U.S. Department of Energy (formally ERDA) has recommended increased research on satellite power, and if that research is conducted effectively by the U.S. National Aeronautics and Space Administration (NASA) and by DOE the question of technical feasibility should be answered during the next few years.[7] The questions of cost and environmental compatibility remain open, but interest is growing rapidly in the potential cost-effectiveness of bringing the necessary materials for the satellites not from inside the biosphere of the Earth, but from known sources outside: the moon and certain Earth-approaching asteroids.

A number of calculations made by economists and technical study groups indicate that this approach to satellite powerplant construction should cut the costs by a large factor, and so bring solar energy into a competitive price range. Recent work also indicates that this unconventional but scientifically well-based approach should permit large-scale production of power satellite components without the need for any rocket vehicle more advanced than the space shuttle already under construction.

Progress has been rapid since the first scientific article, published in December 1975, which suggested that satellite power and space manufacturing might form a winning combination.[5] A NASA study in 1976 explored the most fundamental technical problems of space manufacturing and found no "show stoppers".[3] In 1977, a study more than four times as large was held, going into still greater depth.[4] Participants were sent by NASA laboratories and by private industry. The conclusion was that an Apollo-sized program could achieve the goal of the construction of the first satellite power station from lunar materials by 1991, given a favorable climate of political decision-making.

A Princeton conference in 1975 explored these issues;[8] three additional conferences, on a larger scale and supported by NASA, DOE, and private industry, were held in 1977, 1979, and 1981 in Princeton.[9] Numerous other conferences have been held or are planned. At least six books have been published on the subject of space colonies and satellite power.[10-15] At Princeton we have received over 10,000 inquiries from interested professionals, press, and laymen from a wide variety of disciplines. Progress has been so rapid that a review of the current status of the studies is timely.

II. PROFESSIONAL PARTICIPATION

Individuals from virtually every professional discipline would be involved in achieving the goals of space manufacturing, satellite power, and the long-term human habitation of space. Physicists are required to apply their knowledge of celestial mechanics, electricity and magnetism, thermodynamics, and radiation-shielding to the problems of transport of nonterrestrial materials and the construction of structurally sound and safe habitats, ships of exploration, and solar power stations. Chemists and metallurgists are needed to design the processes of extraction of metals and ceramics for structural materials, glass for windows and solar collectors, oxygen for breathing and an oxidizer for rocket fuel. Biologists and physicians will define the necessary inventories of water, carbon, and nitrogen for consumables, and criteria for human comfort in space, examining the parameters of artificial gravity and angular velocity, the length of stay for the first settlers, and the feasibility and constitution of the first closed ecological systems in space. Engineers will be needed to convert each concept to practical, reliable hardware. Architects will aid in habitat design, and economists will evaluate cost-effectiveness in terms of returned energy to the Earth. The conferences and studies held so far show strong representation from all these fields. As an example, the names of authors and papers at workshops and conferences held during 1976, 1977, and 1979 are shown in the Appendix.

Comparable numbers of social scientists are beginning to address what may turn out to be more difficult questions than the technical feasibility and cost-benefit tradeoff of the proposed program. Is such an undertaking desirable? Who will manage the program? What will be the secondary benefits and costs of the program here on Earth? How will the space environment affect the first settlers? Who will govern? What are the cultural and philosophical implications worldwide? How are existing space treaties affected by the mining of several million tons of lunar and asteroidal material small as those quantities are in comparison with terrestrial mining? What are the international implications of emplacing dozens of large satellite power stations in geostationary (24-hr) Earth orbit? How will the program change our perceptions and policies regarding the use of energy, food, and natural resources on Earth? What new opportunities for human exploration and use of space will arise and how will we take advantage of them? Several of these questions have been addressed at the conferences listed in the Appendix, as indicated by the titles and authors.

Anthropologists, psychologists, psychiatrists, sociologists, philosophers, historians, lawyers, businessmen, political leaders, and writers with a broad interdisciplinary perspective are already addressing these difficult questions, with rigor and foresight. Rarely has an opportunity arisen for collaboration and planning by such a wide variety of individuals. The authors believe potential effect of this undertaking on the progress of human civilization could be substantial.

To assist planners in the early years, the Universities Space Research Association, representing 55 universities, now has a Task Group working in this area. Advising one of the authors (O'Neill), as Chairman of the group, is a panel of experts drawn from many fields: physics, space science, engineering, the humanities, psychiatry, administration of professional organizations, labor unions, and government. Significantly, both the electric utilities industry and major life insurance companies, the prime source of utilities investment, are supplying senior personnel for that panel. In its early meetings, the panel has formulated strategies for early planning, coordination, and timely funding of the most significant and long-range studies and issues. Panel membership and disciplines are listed in Table 1.

III. THE LOW PROFILE ROAD: A SCENARIO FOR THE SPACE MANUFACTURING OF NONTERRESTRIAL MATERIALS

The 1976 and 1977 NASA summer studies, and work carried out in the interval between those efforts[16,17] have defined a scenario for the establishment of the first space manufacturing facility, chemical processing plant, lunar mining operation, and launch device for the lunar materials, based entirely on the Space Shuttle for Earth to Low-Orbit transportation. One major new development for transportation in space is required: the mass driver.

The mass driver is a linear synchronous motor that converts electrical energy into kinetic energy, accelerating 0.001- to 10-kg slugs to high velocities. Each payload-carrying bucket contains superconducting coils and is supported without physical contact by means of dynamic magnetic levitation, a principle known for over 60 years. As in the case of a linear synchronous motor-generator, acceleration is by magnetic fields. In the mass driver, each bucket releases its payload, decelerates with return of energy to the power supply, and picks up another payload for acceleration. The mass driver conversion efficiency from electrical energy to kinetic energy is between 75 and 95%. The power source is either solar or nuclear. A recent model has demonstrated a performance of $30g$ acceleration similar to the value assumed when the mass driver was first published;[18,19] much higher values are theoretically possible and are being planned in the design of the next model.

The mass driver can be used either as a launcher of lunar material into free space or as a reaction engine in space, supplying continuous thrust to transfer large payloads from orbit to orbit in a spiral trajectory.[20,21] In the latter case, aggregates of powder ground from the (otherwise jettisoned) external tanks of the space shuttle would be used as reaction mass, until lunar and asteroidal materials become available. The performance of the mass driver could exceed that of the space shuttle main engines in terms of exhaust velocity or specific impulse. But the mass driver has the advantages that any material can be used as fuel and that solar energy is available in space as the power source.

The studies[17,22] have shown that the requirement for the initial moon-based mining and delivery system employing the mass driver is on the order of 1000 tons on the lunar surface. The mass driver installation implied by that much mass would initially launch to the space manufacturing facility about 30,000 tons/year, but the installation would have the potential capacity of 600,000 tons/year. The initial investment in payloads needed to get to that first level, including propellants about equal in quantity to what has to be landed on the moon, and resupply for personnel in space from the surface of the Earth, would require approximately 100 shuttle flights. This estimate is based on the assumption that personnel are transferred back and forth by the "historical" method—a chemical tug—and that freight can be transferred using a mass driver reaction engine. The investment is relatively modest, therefore; less than 2 years of shuttle flights at the projected traffic model of 50 or 60 flights a year.

At this point there would be no processing or agriculture in space, and a total of only

Table 1
MEMBERS OF USRA COUNCIL ON POWER FROM SPACE

Gerard K. O'Neill (Chairman)
Department of Physics
Princeton University
Princeton, N.J.

James Arnold
Department of Chemistry
University of California, San Diego

John Billingham
Bio-Technology Division
NASA-Ames Research Center
Moffett Field, Calif.

T. Stephen Cheston
Associate Dean, Graduate School
Georgetown University
Washington, D.C.

David R. Criswell
Department of Chemistry
University of California, San Diego

Alexander Dessler
(President, Universities Space
   Research Association)
Department of Space Physics and
   Astronomy
Rice University
Houston, Tex.

Gerald W. Driggers
President, L-5 Society
Birmingham, Ala.

Frederick Ferber
Vice President, Corporate Division
Prudential Insurance Company
Newark, N.J.

S. Robert Hart
Vice President, R & D
Southern Company Services
Birmingham, Ala.

Rene Miller
Chairman
Department of Aeronautics and
   Astronautics
Massachusetts Institute of
   Technology
Cambridge, Mass.

James W. Moyer
Southern California Edison
   Company
Rose Mead, Calif.

Brian O'Leary
90 Westcott Road
Princeton, N.J. 08540

Jay Shurley
University of Oklahoma Health
   Science Center
Norman, Okla

W. J. Usery
Former Secretary of Labor

William Winpisinger
President, International
   Association of Machinists
   and Aerospace Workers

Richard G. Woodbridge, III
2nd Vice President, Investments
New York Life Insurance Company
New York

Jerry Grey
Director of Public Policy
American Institute of Aeronautics
   and Astronautics
New York, N.Y.

about 40 people would be involved. But the achievements would be substantial. First, there would be the proof that the transfer of lunar materials can be done. Second, the initial transport would provide all the cosmic ray shielding needed for the first workforce habitats in space. Finally, the lunar materials would constitute sufficient reaction mass for mass-driver interorbital tug operations in space. A shuttle-derived lift vehicle would be capable of a much higher ratio of payload to the external tankage than the current shuttle, so once the shuttle-derived freight rocket comes into use the shuttle tankage mass (35 tons per flight) would not be sufficient for reaction mass. The lunar soil would be this next source of reaction-mass.

The next step would be to begin processing at the level of 30,000 tons/year, which implies a total output of silicon and metals of some 9000 tons a year. The plant mass required is rather small, 150 tons; the power supplies to drive it are much bigger. If the habitats are designed at 10 tons per person, with 150 people in space needed at that stage, one must put about 1500 tons into the habitat itself. At this second plateau it would be possible to produce relatively simple fabricated products such as silicon solar-cell power arrays and small modular habitats.

In the course of chemical processing, large quantities of oxygen would be generated. Of the 30,000 tons of mass processed per year, photovoltaic arrays would constitute about 1600 tons, habitat components about 7500, and much of the balance would be oxygen. The oxygen could be used three ways: as fuel (propellant oxidizer for the chemical tugs needed to move personnel back and forth in space and as reaction mass for the freight-carrying mass driver), for breathing, and as the heaviest component of water.

At this point production would be increased by adding plants in parallel. The reason for that choice, rather than building bigger plants, is that once the research and development has been done on a small plant, it would be preferable not to repeat it on a bigger one. Small parallel plants would also provide greater reliability. Over a 4-year period, this level of production would provide habitats adequate for some 3000 people. During the same period about a ton per person of organics would be brought in to stock later farming.

Figure 1 shows the history of such a scenario. Zero time represents the start of lift from the Earth. Prior to that, of course, there would have to be substantial research, development, and construction. Figure 1 shows that at the end of 7 years we could build to a throughput of 630,000 tons of lunar material per year. To summarize the economics, one must spend about $20 to $40 billion for research and development, based on space shuttle R&D experience. The transport costs would be a little over a billion dollars per year for 7 years. If the satellite power stations are sold for $1000 per kW at the busbar on Earth, less than the price of a hydroelectric plant today, the earnings would be some $20 billion per year. Total investment to the point of such payback would be roughly equivalent to one Apollo project.

If production is to expand substantially from that point, it will be necessary to use something more advanced than a shuttle-derived lift vehicle. With that vehicle, Figure 2 shows the investment and earnings sequence. With this sort of bootstrap approach, the time horizon for profits drops to 5 to 10 years, not much beyond the range of interest for private investors. Technically it appears that a program of this type could be started somewhere in the mid 1980s. But technical readiness must first be proven. Research funded by the U.S. government has already begun on the most significant components of the program: small NASA grants for mass driver development and for research and development on chemical processing of nonterrestrial materials.

During the 1977 NASA Summer Study, a group of engineers and economists led by Mr. John Shettler of the General Motors Corporation further refined the steps required to manufacture the first satellite power station from nonterrestrial materials. They concluded that this goal could be achieved by 1991 with a total investment of $60 billion dollars (1977), with cost-effective delivery of electricity on the Earth soon afterward.[22] Assumed was a proof-of-concept in the form of pilot plants and experiments on Earth and in space by 1985, a time frame and scope of activity that is within the planned frequency of space shuttle flights by NASA during the early 1980s.

An alternative to the use of lunar resources for space manufacturing is the use of earth-approaching asteroidal materials.[23] O'Leary explored the possibility of using mass driver tugs to move earth-approaching asteroids at opportunities of low velocity

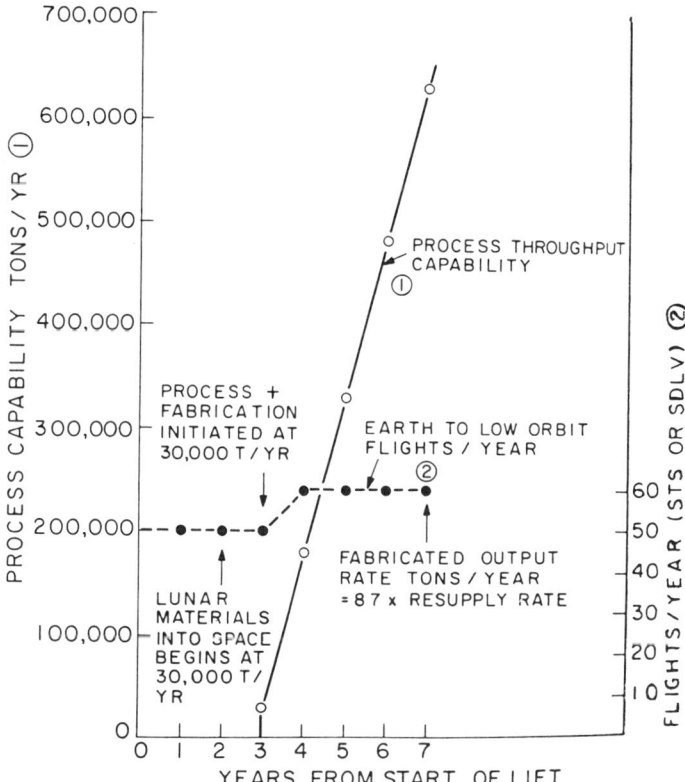

FIGURE 1. Growth of space manufacturing after "lift", defined as that time when the go-ahead has been given to establish a lunar mining operation and nonterrestrial materials processing facility.

increment to the vicinity of the Earth.[24-27] Carbon, hydrogen, nitrogen, and free metals—apparently scarce on the moon—may become available in abundance from some of these objects and it is possible that the retrieval of asteroidal materials may be cost-competitive with that of lunar materials in an early program of space manufacturing. A scenario was developed in which a mass-driver, assembled in space, would retrieve 0.5 to 2 million tons of asteroidal material through a velocity increment of 3 km/sec in about 5 years.[25,26] These velocity increments are comparable to those to and from the lunar surface, but no soft landings would be required and solar landings would be required and solar energy would be continuous at the asteroid. Many such candidate asteroids are believed to be within reach of earth-based telescopes in ongoing search programs.

An asteroid resource study group convened at the NASA-Ames Research Center during the summer of 1977 to redefine the scenario in light of more sophisticated techniques in calculating trajectories using gravity-assist.[25] Two other papers describe in detail the velocity-increment requirements for favorable round-trip missions to currently-known candidates and probable future cases,[28] and an assessment of asteroidal resources and recommendations for expanding the search program, follow up for orbital determination and chemical classification, and identification of precursor missions.[29] A scenario for asteroid retrieval was developed, based on the best known case for a "real" object (asteroid 1977 HB) with gravity-assists from the Earth, Venus, and moon and a likely hypothetical case, given an increase in the asteroid discovery rate

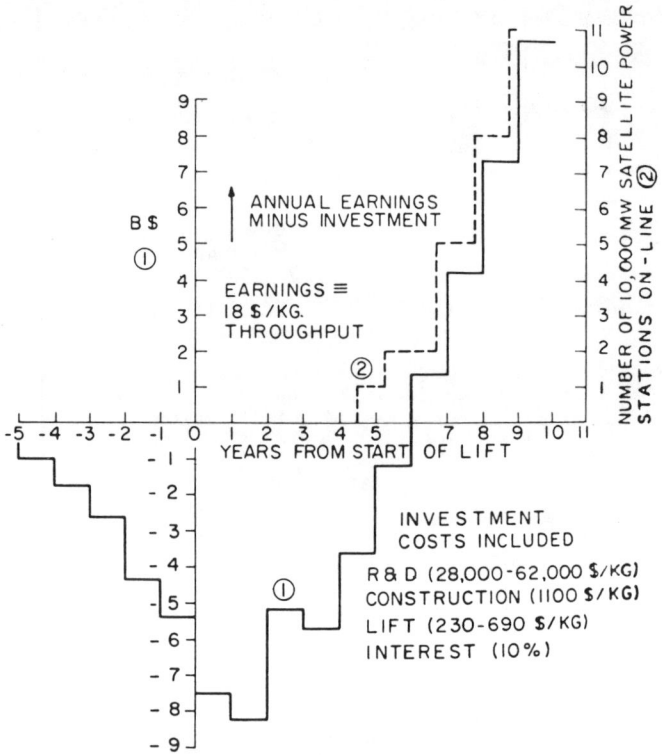

FIGURE 2. Investment and earnings scenario for space manufacturing. Crossover, or "ignition", occurs six years after "lift". (See Figure 1.)

and improved mission analysis techniques. This scenario was then placed in a parametric context, which identified the most significant variables in comparing the economics of transport, into a stable high orbit, of asteroidal, lunar, and terrestrial materials. The study concludes that asteroidal and lunar resources are competitive and are far less expensive to place in high orbit than terrestrial materials. The group recommended a research and development program designed to provide technology readiness for asteroid retrieval by the mid 1980s. Such a program would include an expanded search program and a new start on precursor missions to prime candidates selected for their accessibility and class of chemical composition. These recommendations were corroborated in a second NASA-funded workshop convened later in the summer in La Jolla, Calif.[30]

IV. MINIMAL RISK LUNAR MATERIALS RESEARCH AND DEVELOPMENT PROGRAM

In spite of the promise of bootstrapping a lunar and/or asteroidal satellite power program, the required investment is large enough to make a full program commitment in the near future a challenging task. A key question, therefore, is whether these facilities could be scaled down to a size where concepts of lunar materials processing and transport could be proven and where smaller-scale products could be manufactured in space from lunar materials.

This question of minimal scaling and related technical problems regarding lunar material guidance trajectory, guidance, and mass receiver design were addressed in a series of workshops during 1978 and 1979, sponsored by NASA and the Space Studies Institute. We published the results in the October 1980 issue of *Astronautics and Aeronautics*. This section summarizes the most salient points. Also, many of the technical details of lunar materials transport have been addressed in the published summer studies and Princeton conferences (see Appendix I to this chapter). The results were encouraging: workshop participants found that it would be possible to land relatively small tonnages of equipment on the lunar surface at a low fraction of the investment required for a satellite power program.

Among the questions which the workshop group addressed were the following: What is a minimal facility on the moon geared to processing useful products from lunar materials? Two scenarios are envisioned: one being the definition of a facility on the moon which could begin to process lunar material into a potentially useful inventory (oxygen is most likely); and the second being the definition of a facility which would enable the use of lunar materials in space (e.g. cosmic ray shielding, reaction mass, processed oxygen, or fuel). A related question is: what is the smallest rate of throughput which could accommodate this? What is the most that can be done with one ton landed on the moon? What is the minimal manned facility on the moon? What is the minimal facility capable of exporting material from the lunar surface? Can it be manned? What is the minimal orbiting facility required to process for oxygen? What is the minimal lunar facility required to process for oxygen? What is the minimal orbiting facility required to process for oxygen, silicon, and a metal? What is the minimal lunar facility required to process for oxygen, silicon, and a metal?

The group came to a number of conclusions. First, it would be possible to build up a lunar inventory of oxygen for fuel cells and other products prior to a manned visit. A minimal facility could produce 10 tons/year of stored fuel cell materials on the lunar surface. The power requirement is only about 8 kW. Processed metals and vapor deposited could be used for power supply for later use.

In other words, the answer to the question about the minimal facility on the moon is one capable of producing 10 tons/year of the seven major elements on the moon: oxygen, silicon, iron, aluminum, titanium, magnesium, and calcium. A chemical plant of 200 kg is adequate, so the problem of what can be done with one ton landed on the moon can be solved.

Regarding the question of the evolution of a lunar materials processing program, the group concluded that the masses of both the chemical processing plant and lunar mass driver could be scaled down to a few tons, where the yearly materials throughput-to-mass ratio would be on the order of 100; that is, the mass of both the processing and transport machinery could be replicated in completed products in a span of a few days.

The workshop participants felt that, as a first step, a 1 to 2 ton experimental processing plant be placed in low Earth orbit using simulated lunar soil. One goal would be to obtain products of commercial use, of which oxygen (fuel, fuel cells, and habitat cosmic ray shielding) and silicon (crystals, solar collectors), are major products. The plant would also serve as a demonstration for the later processing of real lunar soil.

The group then investigated two representative scenarios for the development of an early lunar facility. The first would involve the following sequence of steps: (1) land an unmanned 1-ton module on the moon and return it to the low Earth orbit plant for processing of real lunar soil (this will prove the processing experiment and the new lunar transport system at the same time); (2) land an unmanned 1-ton module with a processing plant designed for a 5 year lifetime; it would accumulate an inventory of products for the future (oxygen for fuel cells, breathing and fuel, silicon for solar cells, and metals for structures); (3) land components of a mass driver in 2 1-ton modules capable of exporting 300 tons/year to the space processing plant (which would be either the original small scale one in low Earth orbit or a scaled-up version matched to the higher throughput); immediately land a life support module and ascent stage with six people for a stay of 2 weeks; assemble the mass driver and service the processing plant; start using the lunar inventory of oxygen for fuel cells and breathing; (4) followup with several future landings to upgrade the mass driver by several orders of magnitude to satellite power station scale; also upgrade the space chemical processing plant; one conservative approach would be to do the upgrading in parallel. All steps are separate and the investment for steps 1 and 2 are likely to be less than $1 billion, and step 3 perhaps $5 billion. But the way to bootstrapping could be paved.

The second scenario would involve placing a 2000-ton/year-throughput chemical plant in space, which could fit into one shuttle cargo bay (43 tons). A 30-ton lunar mass driver can launch 2000 tons/year for processing. Allowing for boost to high orbit and a soft lunar landing, about 380 tons would need to be shuttle-launched (13 flights) to establish this lunar facility.

A group consensus developed on some threads common to the small scale lunar materials processing and transport scenarios:

1. Shuttle-launch a pilot processing plant with simulated lunar soil
2. Develop a lunar landing descent/ascent stage
3. Develop an interorbit transfer vehicle
4. Develop a lunar power plant including ways of storing and regenerating and sharing energy
5. Develop a device for collecting lunar material
6. Develop methods of electrostatic and magnetic beneficiation, in addition to ways of crushing into finer sizes

V. ENERGY IMPLICATIONS

There is probably no better example of the difficulty, complexity, and paralysis of research and development policy-making than that of the energy problem. Against the backdrop of debates on the merits and drawbacks of alternative energy sources, rate of growth, environmental vs. industrial interest, regulation, the politics of energy-independence, and the conflicts of planning, there appear obvious dilemmas that are not removed by the current relaxation of the Arab oil embargo. By the year 2000, the world will have run out of the majority of its most well-known natural gas and oil. Alternatives do not appear to provide the answer. Nuclear technology is considered by many individuals as dangerous and may become politically unacceptable; the mining and burning of large quantities of coal are environmentally undesirable; synthetic fuels may not be economically competitive; nuclear fusion is far off in time and may not be economically or environmentally acceptable; and most forms of solar energy are not yet economically competitive and are limited by intermittency, varying local conditions, and a continuing nuclear and fossil-fuel emphasis by most nations. In the U.S. alone, even with the implementation of an increasing conservation ethic, present plans call for more than $1 trillion to be spent on expanding energy use. About half of this cost will be on the capital expansion of central station electrical power generating capacity in coal- and nuclear-fired plants. Meanwhile, the cost of energy continues to increase and future oil embargoes could be crippling.

Most observers think that the long-range energy problem does not have a simple solution. Incremental, patchwork remedies are the natural product of a chaotic decision-making process. Satellite solar power stations which relay energy by microwaves to the Earth may be free of many objections common to other long-range energy options. The technical feasibility of the microwave transmission of energy over a one-mile distance has been established by tests conducted at the JPL Goldstone facility in the Mojave Desert. Separate, laboratory tests have shown an overall efficiency of 55%. Photovoltaic or solar thermal power stations in geosynchronous orbit above the equator have the advantage of supplying any selected area on the Earth with continuous electrical power from a renewal resource. The solar insolation in orbit is between 6 and 15 times greater than the average for fixed solar collectors at favorable locations on the Earth. The power density of the microwave beam would be approximately that of direct sunlight and would be received by a 7-km-diameter antenna array which would provide 10,000 MW at the busbar, the equivalent of ten contemporary nuclear or coal power plants. Current environmental standards for microwave exposure could be met at the perimeter of the receiving station. A few hundred such stations, conveniently located around the world, could supply most of the planet with base-load central-station electricity by 2010 and could provide energy for industries in the countries so that they could increase their standards of living. Terrestrial solar energy (e.g., windmills, collectors, bioconversion, tides, ocean thermal gradients), while more decentralized and usually intermittent, could provide peak loads, and some options for users in some regions. Abundant power would also be available for fertilizer production, for electric vehicles, and for making synthetic fuels.

A DOE task force recently recommended "modest" funding of satellite solar power research and development at a rate of about $5 million per year.[7] The NASA Johnson Spacecraft Center has issued a $1 million request for a proposal to study satellite solar power and has issued a report which suggests that it may be cost-competitive if transportation costs were reduced.[30] More recently, a NASA study on the nonterrestrial materials option has been authorized. Clearly this energy alternative is beginning to be considered as a potentially viable option for widespread use by the turn of the century.

Nevertheless, decision delays have arisen. As in the case of many other solar energy options, many policymakers have made the assumption that because this new technology had been conceived later than existing technologies it must be considered as more esoteric or further off in time. On the contrary, satellite solar power could possibly come on line earlier than the nuclear liquid metal fast breeder reactor, and almost certainly sooner than nuclear fusion. A continuous open mind in assessing priorities will be required to solve the energy problem. Satellite solar power and space manufacturing will need an initial federal R&D impetus before private investment, and that is clearly a matter of public policy and agency responsibility, probably in the U.S.

The main initial objections to satellite solar power were based on the high cost of Earth to high orbit transportation, about $1000/kg using the space shuttle and an upper stage. Current estimates are that 10kg of power station must be emplaced per kW of power provided on the Earth. The resulting 100,000-ton station would cost $100 billion to transport into orbit by the space shuttle system, a prohibitive capital cost when compared to that of competitive energy sources.

New space transportation systems would have to be devised for Earth-launched satellite solar power to become a reality. But if the components of satellite solar power stations were to come from the deep gravity well of the Earth, they would need to be launched by very heavy lift vehicles on a frequent schedule, which would require a new multibillion dollar rocket development program. The recurring transportation costs are uncertain and may remain too high for satellite solar power to achieve economic competitiveness with other renewable energy sources after the development costs are written off. These behemoth rockets could create environmental problems on a scale far exceeding those of the shuttle. The total propellant use would be on the order of 2000 times greater using the big rockets, and there could be problems with a limited propellant supply on Earth, large effluents into the atmosphere,[32] ozone depletion in the stratosphere, sonic booms, and interference with Earth-based astronomy during the low orbit assembly of structures larger than the moon in angular size.

All these economic and environmental objections would appear to be removed by the fabrication of satellite solar power stations from lunar and/or asteroidal materials, using the Shuttle for transferring the necessary equipment. Economic analyses[17,22,33,34] show that the total investment for producing dozens of satellite power stations from lunar materials would be a small fraction of the $800 billion that the electric utilities industry of the U.S. is committed to spending on new generator capacity over the next quarter-century, and indicate that the space alternative should return substantial profits to investors even after the payment of high interest charges.

VI. HUMAN EXPLORATION

Processing more than a million tons of lunar and asteroidal material and constructing and maintaining satellite power stations, each one several square kilometers in area, will inevitably involve the participation of hundreds to thousands of people at the space industrial center. Even on the assumption that all materials for satellite power station construction would come from the Earth, calculations have shown the satisfaction by satellite power of U.S. needs for new generator capacity in 2000 would be most economically carried out by a workforce of several thousand in long-term high-orbital residence.

The first habitats could provide a significant proving ground for living in space: closed ecosystems, agronomy, the physiology and recreation of low and zero gravity, the psychology and sociology of the space community environment. In the long run, the surroundings need not be antiseptic and military like *Star Trek's* Enterprise or the stations of *2001: A Space Odyssey;* rather they could be earthlike with lakes, rivers, mountains, trees, and clouds. A new dimension would open up to the inquiring human spirit and may provide the opportunity to raise human consciousness, perhaps the fulfillment of some of humanities more basic hopes for an open future rich in options. On a more pessimistic note, space settlements could carry the seed of human civilization in the event of nuclear disaster on the Earth.

Beyond the moon remain the mysteries of other planets and the universe. Large scale manufacturing in space would make possible the construction of ships for detailed exploration of the solar system. These analogues to Darwin's Beagle expedition could begin as soon as 1992, the 500th anniversary of the discovery of the New World. Many of the tantalizing scientific questions and findings from the Mariner, Viking, Pioneer, and Venera programs regarding the evolution of planets in the solar system and the origin of life on Earth could be followed up with great care and in depth by such a program. A "hands-on" expedition to the surface of Mars could resolve the ambiguities of the biology experiments on Viking; this could be done 15 to 20 years from now, from a space-manufactured space vehicle, at a small fraction of the $50 to $100 billion which would be required for a manned Mars expedition whose transportation system would come from the surface of the Earth.

Large radio and optical telescope arrays in space could probe the universe and search for extraterrestrial intelligence on a scale far greater than is currently possible. With the entire electromagnetic spectrum opened up, with angular resolutions and apertures increasing orders of mangitude at optical wavelengths, and with the possibility of building large telescope structures that are unimpeded by gravity, astronomy would be revolutionized. A new era of human scientific exploration would be a natural consequence of high orbital manufacturing.

VII. GROWTH AND ENVIRONMENTAL QUALITY

The world stage is set for an ideological battle which could be significant in this generation: should world systems continue to grow or must they settle into a steady state? Growth advocates argue that a rapid transition to a steady state is not possible because the poorer nations are seeking to raise their standards of living to that of the western nations.[33,34] The transition itself, they argue, is contrary to the tenets of existing political systems and would appear to be impossible, on purely practical grounds, for at least decades.

The steady state advocates, on the other hand, feel that growth must stop if humanity is to survive.[37,38] With the limits to growth models taken at face value, even allowing for some room for guessing exactly when they will be reached, the no-growth argument is that political systems and life-styles must adapt to the limits or suffer the consequences which any biological system must when it overpopulates its ecological territory.

Between the extreme views are those which take the world as it is and attempt to solve the obvious problems in a practical way. There is, for example, Benoit's dynamical equilibrium economy[39] which advocates a shift in growth from the depletion of nonrenewable resources to the rapid development of renewable resources, elimination of waste, reduction, and eventual stabilization of population growth, and the reliance on science and technology to provide innovation for services and increased quality of life. Schumacher[40] has proposed that technologies and human activities will need to be developed on a small scale, with individual and small group accountability rather than amorphous large organizations which currently control our destiny. As to whether such measures are politically possible or would be adequate to turn the tide is open to debate.

Regardless of which side of the growth debate someone may wish to take, the space manufacturing concept seems to be a sensible choice for implementation, because satellite power stations would provide a dramatic shift toward the use of renewable energy resources. This component to a world solar energy economy could become the basic underpinning for global stabilization of energy and food use, which may come barely in time to avert massive shortages 20 to 30 years from now in nations like India.[41] It may also be possible that an appreciable share of the metal supply of the world may come from the asteroids[42] and that food grown in space could be delivered cost-effectively to the Earth.[43,44] These fundamental considerations appear to be essential to human survival, whether we emerge as a growth or as a no-growth civilization.

There is one serious possible implication of space manufacturing which will need careful attention in a political context; its use primarily as a mechanism to expand the limits to growth and its implications for continued exploitation. No-growth advocates and ecologists, particularly, have felt this rationale to be undesirable.[45] An exploitation of space which provides mankind a relief valve from the biosphere could result in a cancerous one-way growth into the solar system and would merely buy time before coming up against new limits. Visions are conjured of profit-hungry exploitation, creating energy overdemand and the relaxation of dealing with our problems here on Earth now that we have a panacea.

In other words, the solution to a critical human problem, the shifting the consumption of nonrenewable resources to renewable resources and raising the standard of living of the poor countries, may help to create a new problem—exploitation on a larger scale and relaxation of interim standards. Fortunately the timescales are vastly different. The asteroids contain enough material to provide abundantly for populations tens of thousands of times that of the Earth currently, and solar energy in space is abundant.

These considerations should provide ample warning that an international political framework for controlling the uses of space should be developed. About 100 million tons of material—an excavation on the moon 1 km square and 15 m deep, or a 300-m-diameter asteroid—are about all that is needed to construct enough power satellites to supply the world with its energy in 2000. Eventually it may be desirable to establish large colonies for developing exploratory opportunities and alternative lifestyles. In any case, the possibility of a desperate situation here on Earth in the relative near term, and the best hopes of mankind, require a vigorous effort in establishing something like Benoit's dynamic equilibrium economy and the space manufacturing of satellite power stations. To hestitate on the basis of vague fears of runaway growth seems to us unwise, considering that the resources of space are many thousands of times larger than those of the Earth.

VIII. RESEARCH RECOMMENDATIONS

There are several research areas in which a relatively small research investment could have potentially large payback:

1. Silicon purification from nonterrestrial materials, for solar cell fabrication
2. The design of satellite power stations optimized for the use of nonterrestrial materials
3. Development of the mass driver concept, through a sequence of working models, to an orbital test engine brought up as a shuttle payload
4. Chemical processing of minerals similar to the lunar soils, in a sequence of bench-top and then pilot-plant experiments
5. Studies of the scaling of pilot plants, which affects the question of boot-strapping, and also affects the minimum investment necessary to obtain economic payback from the use of nonterrestrial materials
6. Work-force productivity studies (tons per person-year) that affect the necessary rate of habitat construction, and so may limit the buildup rate of manufacturing in space
7. Modular habitat design, combining efficient modular designs with pleasant aesthetics
8. Environmental impact studies—perhaps the first of these should be the microwave effect on the ionosphere; second is the issue of launch-vehicle emissions. Apparently the existence of a space manufacturing facility would reduce the necessary emissions into the upper atmosphere by a factor of 100 to 1000 as compared with using conventional launches from the surface of the Earth, for equal tons per year of finished products in space.
9. Resource studies of at least three kinds: (a) Radar search for materials that may be trapped in the L4 or L5 Lagrange regions of the Earth-moon system. It is not ruled out that there may be meter-size chunks of materials with an aggregate mass of hundreds of thousands of tons in those locations. (b) A search for Earth-approaching asteroids, using a moderate-to-large aperture Schmidt telescope on the Earth or in orbit. Identify and plan precursor missions to candidate asteroids during the mid 1980's to determine chemical and physical properties. (c) A lunar polar orbiter, previously planned by NASA, but deleted in its 1978 budget.
10. Materials studies using nonterrestrial materials and the space environment; for example, the substitution of silicon, oxygen, and metal composites for the plastic materials that would otherwise have to come up from the Earth; composites such as silica fibers in a soft aluminum matrix. Special fabrication techniques using the space environment, such as metal evaporation in vacuum, a possible simple method for achieving very high productivity in the fabrication of large structures.

11. Research directly related to life-support: e.g., controlled-environment agriculture.

IX. THE CHALLENGE

Both in the political and the business worlds there is little significant planning beyond 3 to 5 years. Therefore it is not practical to attempt forecasting the future even as far as the year 2000. For us to design for 2028 would be as pointless and futile as it would have been for the Wright Brothers, in 1902, before the first powered flight, to have attempted to design aircraft for 1952, by which time nonstop transcontinental air commerce and supersonic jet flight were commonplace. If our work is to be of value, it must rather be based on the near-term hardware which we can understand and evaluate. In that work we should keep in mind that in September 1957, even Sputnik I had not yet flown, while by July 1969, less than 12 years later, two men had made a safe and successful round trip to the lunar surface. On time scales greater than a generation, trends, societal frameworks, and philosophy can be examined, but specific technologies are impossible to predict.

On the other hand, technology planning (as opposed to the unrealizable political planning) for 10, 20, and 30 years from now is worthwhile. We believe that the potential benefit is too large for us not to explore the space manufacturing option. Regardless of the outcome of debates on growth, energy policy, and political philosophy, the outcome of research work done so far is very positive; its promise is sufficiently large that a modest effort toward the development of satellite solar power could properly begin immediately. Unless other new energy technologies become economically practical, the only alternative to development of satellite power appears inevitable and undesirable: growing international inequities in energy and food supply, future oil embargoes, expensive energy, the depletion of nonrenewable resources, the proliferation of dangerous nuclear technology, and possibly massive hunger and starvation.

Starting a logical progression of the steps that would be required if we are to use the vast material and energy resources available to us in space has been difficult, but those first steps have now been taken. Realistically, we must expect that we must carry on intense activities in the areas of education and political negotiation, if the necessary research work is to continue and expand. The research recommendations listed in this chapter are a beginning. The challenge is more one of communication and the slow acceptance of new ideas, rather than technical feasibility and economics. History suggests that it is almost inevitable that this project will go ahead sooner or later. Those of us working toward its realization feel that in view of the age-old desires of humanity for freedom and abundance, our efforts are well spent.

REFERENCES

1. **O'Neill, G. K.** The colonization of space, *Phys. Today,* 27, 32, 1974.
2. **Johnson, R. D. and Holbrow, C., Eds.,** *Space Settlements: A Design Study,* NASA SP-413, National Aeronautics and Space Administration, Washington, D.C., 1977.
3. **O'Neill, G. K. and O'Leary, B. T., Eds.,** *Space Manufacturing from Nonterrestrial Materials,* Vol. 57, American Institute of Aeronautics and Astronautics, New York, 1978.
4. **Billingham, Gilbreath, W. and O'Leary, B., Eds.,** *Space Resources and Space Settlements,* NASA SP-48, National Aeronautics and Space Administration, Washington, D.C., 1979.
5. **O'Neill, G. K.,** Space colonies and energy supply to the Earth, *Science,* 190, 943, 1975.
6. **Glaser, P. E.,** Space shuttle payloads, hearing before the Committee on Aeronautical and Space Sciences, U.S. Senate, 93rd Congress, 1st Session on Candidate Missions for the Space Shuttle, part 2, Government Printing Office, Washington, D.C., October 31, 1973, 11.

7. Final Report of the ERDA Task Group on Satellite Power Stations, Office of the Adminstrator, Energy Research and Development Administration (now DOE), Washington, D.C., November 1976.

8. **Grey, J. Ed.,** *Space Manufacturing Facilities,* Proc. 1974, 1975 Princeton Conferences, American Institute of Aeronautics and Astronautics, New York, 1977.

9. **Grey, J. Ed.,** *Space Manufacturing Facilities II,* Proc. 1977 Princeton Conference, American Institute of Aeronautics and Astronautics, New York, 1977; Grey, J., Ed., *Space Manufacturing Facilities III* (Proc. 1979 Princeton Conference), American Institute of Aeronautics and Astronautics, New York, 1980.

10. **O'Neill, G. K.,** *The High Frontier,* William Morrow and Company, New York, 1977.

11. **Heppenheimer, T. A.,** *Colonies in Space,* Stackpole Books, Harrisburg, Pa., 1977.

12. **Golden, Frederic,** *Colonies in Space,* Harcourt Brace Jovanovich, New York, 1977.

13. **Knight, D. C.,** *Colonies in Orbit,* Wiliam Morrow and Company, New York, 1977.

14. **Brand, S.,** *Space Colonies,* Penguin Books, New York, 1977.

15. **Avery, N.,** *Time Out for Tomorrow,* T. H. A. R. Institute, Raynesford, Montana, 1977.

16. **O'Neill, G. K.,** Engineering a space manufacturing center, *Astronaut. Aeronaut.,* 14, 20, 1976.

17. **O'Neill, G. K.,** The low (profile) road to space manufacturing, *Astronaut. and Aeronaut.,* 16, 24, 1978.

18. **Kolm, H. H.,** Basic coaxial mass driver reference dsesign, in *Space Manufacturing Facilities II,* Grey, J., Ed., American Institute of Aeronautics and Astronautics, New York, 1977, 91.

19. **Fine, K.,** Basic coaxial mass driver construction and testing, in *Space Manufacturing Facilities II,* Grey, J., Ed., American Institute of Aeronautics and Astronautics, New York, 1977, 103.

20. **O'Neill, G. K.,** Mass driver reaction engine as shuttle upper stage, in *Space Manufacturing Facilities II,* Grey, J., Ed., American Institute of Aeronautics and Astronautics, New York, 1977, 109.

21. **O'Neill, G. K. and Kolm, H. H.,** Mass driver for lunar transport and as a reaction engine, presented at International Astronautical Federation XXVIIIth Congress, Prague, Czechoslovakia, September 25 to October 1, 1977.

22. **Vajk, J.P., Engel, J. H., and Shettler, J. A.,** Habitat and logistic support requirements for the initiation of a space manufacturing enterprise, in Space Resources and Space Settlements, Billingham, J., Gilbreath, W., and O'Leary, B., Eds., NASA SP-428, National Aeronautics and Space Administration, Washington, D.C., 1979.

23. **O'Leary, B.,** Mining the Apollo and Amor asteroids, *Science,* 197, 363, 1977.

24. **O'Leary, B.,** Mass driver retrieval of Earth-approaching asteroids, in *Space Manufacturing Facilities II,* Grey, J., Ed., American Institute of Aeronautics and Astronautics, New York, 1977, 157.

25. **O'Leary, B., Gaffey, M. J., Ross, D. J., and Salkeld, R.,** The retrieval of asteroidal materials, in *Space Resources and Space Settlements,* Billingham, J., Gilbreath, W. and O'Leary, B., Eds., NASA SP-428, National Aeronautics and Space Administration, Washington, D.C., 1979.

26. **O'Leary, B.,** Asteroidal resources for space manufacturing, presented at International Astronautical Federation XXVIIIth Congress, Prague, Czechoslovakia, September 25 to October 1, 1977.

27. **O'Leary, B.,** Resource potentials of asteroid capture, in *Macro-Engineering and the Infrastructure of Tomorrow,* Davidson, F. P., Giacoletto, L. J., and Salkeld, R., Eds., American Association for the Advancement of Science, Washington, D.C., 1978, 209.

28. **Bender, D. F., Dunbar, R. S., and Ross, D. J.,** Round-trip missions to low-delta-v asteroids and implications for material retrieval, in Space Resources and Space Settlements, Billingham, J., Gilbreath, W., and O'Leary, B., Eds., NASA SP-428, National Aeronautics and Space Administration, Washington, D.C., 1979.

29. **Gaffey, M. J., Helin, E. F., and O'Leary, B.,** An assessment of near-Earth asteroid resources, in Space Resources and Space Settlements, Billingham, J., Gilbreath, W., and O'Leary, B., Eds., NASA SP-428, National Aeronautics and Space Administration, Washington, D.C., 1979.

30. **Arnold, J. R. Ed.,** Summer Workshop on Near-Earth Resources, NASA Conference Publication 2031, Washington, D.C., 1977.

31. Solar Power Satellite: Concept Evaluation, NASA Report JSC-12973, Johnson Space Center, Houston, 1977.

32. **Chase, R. L.,** Satellite Power Stations (SPS): Environmental and Health Impact, General Research Corporation, McLean, Va., May 1977.

33. **Hopkins, M. M.,** Cost-benefit analysis of space manufacturing facilities, in *Space Manufacturing Facilities II,* Grey, J., Ed., American Institute of Aeronautics and Astronautics, New York, 1977, 305.

34. **Driggers, G. W.,** A factory concept for processing and manufacturing with lunar material, in *Space Manufacturing Facilities II,* Grey, J., Ed., American Institute of Aeronautics and Astronautics, New York, 1977, 183.

35. **Heilbroner, R.,** *An Inquiry Into the Human Prospect,* W. W. Norton, New York, 1974.

36. **Rustin, B.,** No growth has to mean less is less, in *New York Times Magazine,* May 2, 1976.

37. **Daly, H. E.,** *Toward a Steady-State Economy,* H. H. Freeman and Company, San Francisco, 1973.

38. **Meadows, D. H.,** *et al., The Limits to Growth,* Universe Books, New York, 1972.

39. **Benoit, E.,** The coming age of shortages, *Bull. At. Sci.,* January, February, and March, 1976.

40. **Shumacher, E. F.,** *Small is Beautiful,* Perennial Library, New York, 1975.

41. **Mayur, R.,** Solar Energy, lecture presented to the Forum of Free Enterprise, Bombay, July 3, 1978.

42. **Gaffey, M. J. and McCord, T.,** Mining outer space, *Technol. Rev.,* 79, 50, 1977.

43. **O'Leary, B.,** Food and raw material supply from space to the Earth, in *Escasez Mundial de Alimentos y Materias Primas,* Vicuña, F. O., Ed., Univ. de Chile, Santiago, 1977.

44. **O'Leary, B.,** Limits to Growth Implications of Space Settlement, presented at AAAS Symp. Prospects for Life in the Universe: The Ultimate Limits to Growth, Washington, February 12 to 17, 1978.

45. See, for example, the Spring 1976 issue of *The Co-Evolution Quarterly,* Sausalito, Calif.

APPENDIX

1. "Space-Based Manufacturing from Nonterrestrial Materials, in *Progress in Astronautics and Aeronautics,* Vol. 57, Vol. Eds. **O'Neill, G. K.** and **O'Leary, B.,** Series Ed. Summerfield, M., AIAA, 1977—Technical papers derived from the 1976 Summer Study at NASA Ames Research Center, Moffett Field, Calif.
   O'Neill, G. K., The Concept of Space-Based Manufacturing Facilities; **O'Leary, B., Heppenheimer, T. A.,** and **Kaplan, D.,** Trajectory Analyses for Material Transfer from the Moon to a Space Manufacturing Facility. **Chilton, F., Hibbs, B., Kolm, H., O'Neill, G. K.,** and **Phillips, J.,** Electromagnetic Mass Drivers; **Chilton, F., Hibbs, B., Kolm, H., O'Neill, G. K.,** and **Phillips, J.,** Mass-Driver Applications; **Phinney, W. C., Criswell, D., Crexler, E.,** and **Garmirian, J.,** Lunar Resources and Their Utilization; **Criswell, D.,** Appendix: Materials Packaging; **Driggers, G. W.** and **Newman, J. E.,** Establishment of a Space-Manufacturing Facility, **O'Neill, G. K.,** Appendix: Maximum-Strength, Minimum-Mass Structures; **O'Neill, G. K.,** and **Driggers, G. W.,** Appendix: Observable Effects and Human Adaptation to Rotating Environments.

2. Technical papers derived from the 1977 Summer Study at NASA Ames Research Center, Moffett Field, Calif., NASA publication (in press):
   Spurlock, J. M., and **Modell, M.,** Systems Engineering Overview for Regenerative Life-Support Systems Applicable to Space Habitats; **Spurlock, J. M.,** et al, Research Planning Criteria for Regenerative Life-Support Systems Applicable to Space Habitats; **Bock, E. H., Lambrou, F., Jr.,** and **Simon, M.,** Effect of Environmental Parameters on Habitat Structural Weight and Cost; **Vajk, J. P., Engel, J. H.,** and **Shettler, J.,** Habitat and Logistic Support Requirements for the Initiation of a Space Manufacturing Enterprise; **O'Neill, G. K.,** Mass Drivers, I: Electrical Design; **Bowen, S.,** Mass Drivers, II: Structural Dynamics; **Kolm, H.,** Mass Drivers III: Engineering; **Bender, D. F., Dunbar, R.S.,** and **Ross, D. J.,** Round-Trip Missions to Low-Delta-V Asteroids and Implications for Material Retrieval; **O'Leary, B., Gaffy, M. J., Ross, D. J.,** and **Salkeld, R.,** Retrieval of Asteroidal Materials; **Gaffey, M. J., Helin, E. F.,** and **O'Leary, B.,** An Assessment of Near-Earth Asteroid Resources; **Criswell, D. R.,** Initial Lunar Supply Base: Resources and Construction; **Ho, D.** and **Sobon, L. E.,** Extraterrestrial Fiberglass Production Using Solar Energy; **Lee, S. M.,** Lunar Building Materials—Some Considerations on the Use of Inorganic Polymers; **McKay, D. S.,** and **Williams, R. J.,** A Geologic Assessment of Lunar Ores; **Rao, D. B., Choudary, U. V., Erstfeld, T. E., Williams, R. J.,** and **Chang, Y. A.,** Extraction Processes for the Production of Aluminum, Titanium, Iron, Magnesium, and Oxygen from Non-terrestrial Sources; **Williams, R. J., McKay, D. S., Giles, D.,** and **Bunch, T. E.,** Mining and Beneficiation of Lunar Ores.

3. 28th International Astronautical Congress, International Astronautical Federation (IAF), Prague, Czechoslovakia, September 25 to October 1, 1977, Symposium on Space-Based Industry: Extra-Terrestrial Mining and Delivery Systems:
   Driggers, G. R., Systems Analysis of Space-Manufacturing (77-72); **Criswell, D.,** Packaging of Lunar Materials for Transport (77-73); **O'Neill, G. K.** and **Kolm, H.,** Mass Driver: For Lunar Transport and As A Reaction Engine (77-74); **Leonovich, A. K., Gromov, V. V., Semyonov, P. S., Surkov, Y. A., Permiov, V. G.,** and **Dmitriev, A. D.,** On the Physical and Mechanical Properties of Venusian Soil (77-75), **Alexandrov, A. K., Borisova, I. B., Gromov, V. V., Zakharov, V. V.,** et al, Automatic Devices for Studying the Physical and Mechanical Properties of Planetary Soil (77-76); **O'Leary, B.,** Asteroidal Resources for Space Manufacturing (77-77); **Heiss, K. P.,** Economics of Non-Terrestrial Mining and Delivery Systems (77-78).

4. American Astronautical Society, Annual Meeting and Symposium, San Francisco, Calif., October 18 to 20, 1977:

Sperber, R. and Zipursky, H., Freedoms and Constraints in Solar Power Satellite Designs (77-200); **Engler, E. E.,** Automated Space Fabrication of Structural Elements (77-210); **Nathan, A.,** Camparison Analysis of Space Construction Bases (77-202); **Miller, K. H.,** Solar Power Satellite Construction Concepts (77-203); **Stine, G. H.,** Governmental and Industrial Roles in Initiation of Space Industrialization (77-210); **Grodzka, P. G.,** Expanding NASA's Charter to Facilitate Space Utilization (77-211); **Sviedrys, R.,** Energy Crisis: A History Lesson (77-212); **DeMandel, S. L.,** Two Lessons from the Past: An Analysis of Government's Role in Developing Super-Economics (77-213); **Katz, E.,** Structural and Assembly Concepts for Large Erectable Space Systems (77-205); **Vajk, J. P.,** Space Industrialization and the Long-Term Prospects for Terrestrial Civilization (77-226); **Sanders, T. W.,** A Free Enterprise Model for the Operation of Industrial Facilities in Space (77-229); **Gould, C.,** Space Industrialization—The Long-Range View and the Near and Intermediate Steps (77-230); **Stine, G. H.,** Marketing Techniques and Space Industrialization (77-232); **Bell, M. W. J.,** Advanced Launch Vehicle and Technologies (77-217); **McCoy, H. E.,** Ground Operation and Concepts for Future Space Activities (77-218); **Kingsbury, D.,** A Hybrid Chemical-Nuclear Space Frieghter Concept (77-219); **Downey, P.,** The Interim Upper Stage (77-220); **Austin, R. E.,** Solar Electric Propulsion and Inter-Orbital Transportation (77-221); **Moravec, H.,** A Non-Synchronous Orbital Skyhook (77-223); **Moulton, P. M.,** Use of Outer Planet Satellites and Asteroids as Sources of Raw Materials for Life Support Systems (77-236); **Schweitzer, K. K., Wortmann, J., Roszmann, A.,** and **Betz, W.,** Space Processing of Turbine Blades by Means of Skin Technology (77-224); **Bloom, H.,** A Baseline of Logistic and Power Requirements for Full-Scale Manufacturing of Metal Products (77-237); **Hammel, R. L. and Waltz, D. M.,** Roadmap to Space Products (77-239); **Akins, F. W.,** Isolation and Confinement: Considerations for Colonization (77-245); **Copeland, J. W.,** Communications Requirements of a Space Settler: The Need to Keep in Touch (77-246); **Cutler, W. H.,** Personal Growth Education for Space Colony Inhabitants (77-247); **Conley, C.,** Women's Future in Space (77-248); **Leary, T.,** The Physchological Effects of High Orbital Migration (77-249); **Cepin, M. M.,** Analysis of a Space Community: Possible Areas of Deviance (77-250); **David, L.,** Military Uses of Outer Space (77-251); **Urbanowicz, C. F.,** Cultural Implications of Extraterrestrial Contact and the Colonization of Space (77-252); **Billman, K., Bowen, S., and Gilbreath, W.,** Satellite Mirror Systems for Providing Terrestrial Power: System Concept (77-240); **Muller, R.,** In Orbit Manufacture of Solar Reflector Satellites (77-241); **Clark, L. C., DiBattista, J., and Huckins, III, E. K.,** LDEF/Shuttle Capabilities for Environmental Testing in Space (77-233); **Stevens, N. J., Purvis, C. K., and Berkopec, F. D.,** Interaction of Large, High-Power Systems with the Operational Orbit Charged Particle Environment (77-243); **Prehoda, R. W.,** The Inevitability of Extra-terrestrial Robotics in Space Industrialization (77-244); **Matelan, M. N., and Matelan, L. W.,** The Integrated Use of Space Factories (77-214); **Wolff, E. A.,** Public Service Satellite Communications (77-256); **Ehrlich, M. J., and Zylius, F. A.,** Assembly in Space of Large Communication Structures (77-259); **Lipke, D. W.,** Maritime Satellite Communications, Where We Are and Where We are Going (77-257); **DeSaussure, H.,** The Necessary Elaboration of Space Law for the Commercial Use of Outer Space (77-265); **Frazier, M.,** An International Space Launch Center (77-265); **Menter, M.,** The Impact of Treaties on Commercial Space Operations (77-262); **Robinson, G. S.,** The Outer Space Treaty and the Continuing Military Occupation of Space (77-270); **Rosenfield, S. B.,** The Common Heritage of Mankind' Doctrine and Private Industrial Development of Outer Space (77-269); **Salmon, J.D.,** "Politics of Law for Space Industrialization" (77-271); **Smith, D. D. and Beatty, F. K.,** Communications Via Satellite: A Vision in Retrospect (77-267); **Bock, E.,** Habitats in Space—An Update (77-272); **Spurlock, J.,** Technology Requirements for Closed-Ecology Life Support Systems Applicable to Space Habitats (77-273); **Paluszek, M.,** "Magnetic Radiation Shielding for Permanent Space Habitats (77-274); **Schultz, B. E.,** Space Habitats at the Earth-Moon Lagrange Point (77-275); **Alvarado, U.,** Requirements of Orbiting Facilities for Industrial Space Processing (77-276); **Fedor, O.,** Concept for a Lunar Orbital Logistics Support Station Training Facility (77-277); **Frentz, B., Romani, N., and Mead, G.,** If We Can Get Along at Idaho State, We Can Get Along in Space (77-281); **Motts, C. J.,** Cultural Futuristics: The New Role of Antropology on Earth and In Outer Space (77-292); **Maruyama, M.,** Design Principles and Cultures (77-282); **Bluth, B. J.,** Alternative Social Structures in Vacuum (77-283); **Ziegler, W.,** Weather or Not-Meteorology in Space Cylinder (77-284). **Stuart, M. L.,** Aesthetic Implications of the Crystal Palace Space Habitat (77-285); **MacCallum, S.,** Entrepreneurial Opportunity in the Provision of Community Services in Space Colonization (77-286); **Rudoff, A.,** Public Perceptions of the Space Community Program (77-288); **Bjornen, E.,** How You Gonna Get Them Off the Farm After They've Seen 'Star Wars'? (77-291); **Heppenheimer, T. A.,** Space Community Planning From a Viewpoint of Experience (77-290); **Basler, C.,** International Satellite Staging (77-290).

5. Fourth Princeton Conference on Space Manufacturing, Princeton, N.J., May 14 to 17, 1979:
Michaud, M., Four Dimensional Strategy (79-1370); **Mayur, R.,** Space Manufacturing and theThird World (79-1371); **Pikus, I.,** Status of the Discussions on the Moon Treaty (79-1372); **Dupas, A.,** The Potential Global Market in 2025 for Satellite Solar Power Stations (79-1373); **Finch, E.,** Current Space Habitat Legal Developments (79-1374); **Williams, R., and Erstfeld, T.,** Carbon Dioxide Electrolysis Using a Cermanic Electrolyte (79-1375); **J.,** Solar Power Satellite Beam Disturbance of the Ionosphere (79-1422); **Ziegler, W.,** Atmospheric Attenuation of Centimeter Microwaves (79-1423); **Heppenheimer, T. A.,** Guidance Trajectory and Capture of Lunar Materials (79-1424); **Malzbender, R.,** Optical Scaning of Moving Payload Positions (79-1425); **Von Herzen, B.,** Light Pressure and Solar Wind Perturbations to Payload Trajectories (79-1426); **O' Donnell, C.,** Design Opportunities—Zero Gravity Versus One Gravity Environments (79-1427); **Stuart, M.,** Aesthetic Considerations in Bernal Sphere Design (79-1428); **Thomas, D.,** The Value of Anthropology for Space Settlements (79-1429); **Bluth, B. J.,** Consciousness Alteration in Space (79-1430); **Sterns, P. and Tennen, L.,** The Art of Living in Space: a Preliminary Study for the Local Government of a Space Communty (79-1431); **O'Leary, B.,** Asteroid Prospecting and Retrieval (79-1432); **Ross, D.,** Low Thrust Alteration of Asteroidal Orbit (79-1433); **Singer, C.,** Collisional Orbital Change of Asteroidal Materials (79-1434); **Shoemaker, E., Helin, E., and Bus, S.,** The Search for Earth-Approaching Asteroids (79-1436); **Dunbar, R. S.,** The Search for Asteroids in the L_4 and L_5 Libration Points in the Earth-Sun system (79-1437); **Salkeld, R.,** Parametric Analysis of the Comparative Cost of Recovering Terrestrial Lunar and Asteroidal Materials (79-1438).

Chapter 5

MATERIALS PROCESSING IN SPACE

Robert D. Waldron and David R. Criswell

TABLE OF CONTENTS

I. INTRODUCTION

Industrial activity involving processing of material resources into tangible products has hitherto been confined to regions close to the surface of the earth. With the demonstrable capability to transport and support humans in space orbits and the lunar surface established, it is logical to assume that industrial activity involving material products will eventually extend to various extraterrestrial locations. The cost effectiveness of such extraterrestrial industry will depend on technology development in such areas as transportation systems, mining and beneficiation, materials processing, manufacturing and fabrication, assembly, maintenance, and life support.

A. Classification

Various industrial operations may be classified according to location into 4 regions: Earth, near Earth orbit, far space, and lunar surface. Eventually extensions to other planets such as Mars may become important. The sequence of operations necessary to convert raw materials into useful products may be conducted entirely in one of the specified regions, or materials or semifinished goods may be transferred from one region to another at various stages. Some possible optional routes are shown in Figure 1.

This discussion will be limited to operations falling within the Processing-Refining steps for all locations except the earth. Excluded from Processing-Refining will be those steps comprising mining or gathering of raw materials and physical beneficiation by magnetic, electrostatic, thermal, or combined treatments designed to beneficiate or upgrade raw materials by selective separations. Also excluded will be those post-refining steps primarily concerned with changes in particle size, state of aggregation, geometrical arrangement, or macro- or micro-shapes of materials assemblies which we shall regard as manufacturing operations. Also excluded are those volumetric or surface treatments performed on partially or completely shaped components such as heat treatment, surface hardening, plating, anodizing, or other coating operations, regardless of whether or not chemical modifications result.

Materials processing in space will thus be restricted to operations required to separate desired components from various grades of raw or physically beneficiated input ores, semi-purified feedstocks, or process scrap and convert and/or refine them to elements, alloys, or compounds suitable for subsequent manufacture into finished goods or consumables.

B. Development of Options

Of the various alternative materials flow routes shown in Figure 1, four options have been studied in some detail to determine technical and economic feasibility. These are

1. Earth (mining/beneficiation and/or partial processing)—Space (processing and/or partial manufacturing)—Earth (balance + use)

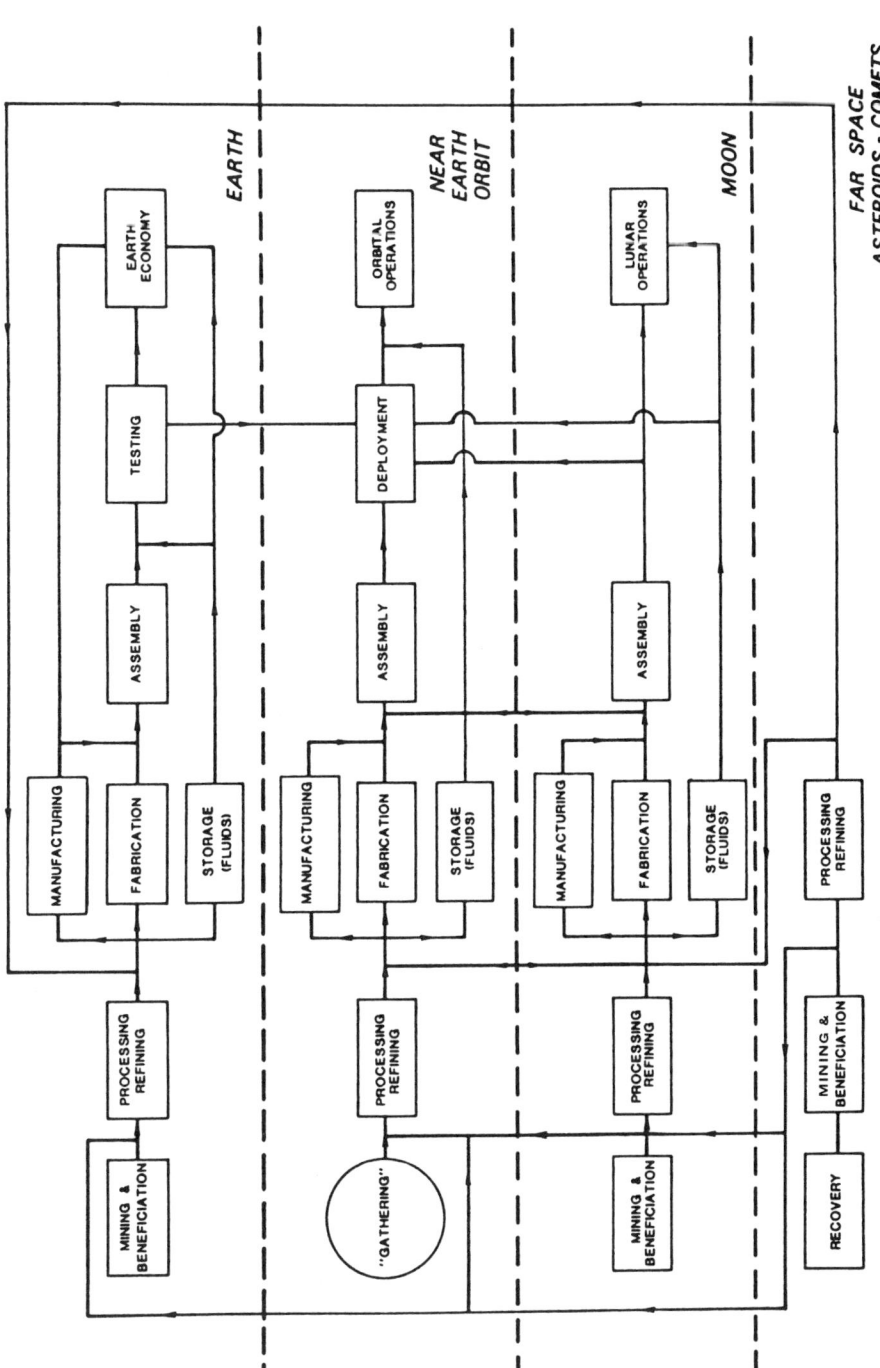

FIGURE 1.   Materials flow options. (From NASA, Lyndon B. Johnson Space Center, Houston.)

Table 1
CHARACTERISTICS OF MATERIALS PROCESSING OPTIONS

Option	Transportation costs	Element constraints	Market size	Competitive costs
1	High	None	Large	Poor
2	High	None	Small	Fair
3	Low	Lunar recoverable elements + small additions	Large	Fair
4	Low	Lunar recoverable elements + small additions	Small	Excellent

2. Earth (mining/beneficiation and/or partial processing)—Space (balance + use)
3. Lunar (mining/beneficiation and/or partial processing)—Space (processing and/or partial manufacturing)—Earth (balance + use)
4. Lunar (mining/beneficiation and/or partial processing)—Space (balance + use)

These options have process characteristics summarized in Table 1.

In point of time option 4 may be expected to become cost-effective first followed by Options 2 and 3 for limited product areas and finally by Option 1 in highly specialized technologies.

Options 1 and 3 must compete with industrial activity completely confined to Earth. To overcome the adverse transportation costs we must achieve substantial economies in Processing-Refining or Manufacturing in comparison with Earth plant practice. This is only likely to occur as a result of the unique properties of the space environment. Of these, the most important is expected to be extended duration of low or zero gravity processing and, to a lesser extent, unlimited volume vacuum processing with vanishing moisture or oxidation problems.

Option 2 must compete with Option 4 or a sequence involving Earth (mining/beneficiation, processing and/or partial manufacture—SPACE [balance + use]). In this case, Option 4 may be expected to display a transportation cost advantage, but the alternate earth starting route traditionally used for space projects has little or no difference in transportation costs from Option 2, and the competitive edge will depend on Processing-Refining and Manufacturing efficiencies. Here again, we may expect that Option 2 will be cost effective when substantial economies in Processing-Refining can be achieved as a result of unique features of the space environment.

II. MATERIALS PROCESS SELECTION

The nature, complexity, and criteria for selection of materials-processing and refining systems may be expected to be substantially different for space processing of Earth-derived vs. lunar or space-derived materials. In this context, scrap materials already in space will be considered as space-derived materials regardless of original source, since their transportation costs would be similar to other space-derived materials.

For the former, the specific output (output mass rate per unit plant mass), energy efficiency, requirements for reagent recycling and element constraints are of little or no importance and a given (specialty) plant may perform a single or small number of operations to produce a small quantity or specific rate of very high specific value product (dollar per kilogram).

Materials processing plants for lunar or asteroidal materials on the other hand will probably be complex, high specific output plants similar to basic industries on Earth, and in which efficient recycling of reagents containing lunar deficient elements must be practiced to avoid excessive mass rates for reagent replacement. After a mature level of space industrialization occurs, a variety of special processing facilities for small output, high specific value products may be established.

A. The Availability of Materials from Space

Table 2 shows the composition of the major constituents of the mare and highlands regions of the moon.[1] Meteoritic bombardment has tended to homogenize the distribution of minerals in these two regions to great depths. There are no aqueous processes operating on the moon to concentrate minerals or elements. Thus the authors expect to work with the dust and surface rocks of the moon, rather than look for deep veins of minerals. Table 3 presents compositional information derived from remote observation of the surfaces of major asteroids.[2,3] It is evident that the chemical/metallurgical industry in space will be substantially different from that on Earth, due to scarcity of key elements such as H, C, Na, Cl, etc. One may also recover gaseous constituents from the atmospheres of Earth or other planets.

In common with industry on Earth, one may anticipate commercial use of both native lunar mineral products (raw or beneficiated) and processed or refined materials (metals, oxides, etc.) for various applications, with price-performance criteria determining use patterns. Native lunar soils may be sintered or fused to obtain a variety of ceramic, cast basalt, and dark-glass products. Free iron may also be recovered by magnetic methods from lunar soil.

Table 4 shows elements that are potentially recoverable from the moon. (The designations "major," "minor," and "trace" are the authors'.) The light trace-elements are mostly due to solar wind bombardment of the lunar surface. Alpha radiation, due to radioactive decay, is responsible for some helium, while impacts of carbonaceous meteorites are responsible for some of the carbon that is present.

The major elements can be recovered using hydrochemical, pyrochemical, electrochemical, or physical processes. This can be performed in orbit or on the moon. The chemical plant would also have to be responsible for the recycling of nonlunar materials. The minor elements could either be corecovered from the major-element processing or be obtained by separate means. In the latter case, the processing could only be performed on the moon, as it would be uneconomical to ship large quantities of soil into space solely for the purpose of obtaining minor constituents.

While the overall abundances of many of the trace elements on the moon do not greatly differ from those on earth, the absence of known concentrated deposits (ore bodies) of such elements makes prospects for their efficient recovery rather dim. Of course, one may still bring critical materials from the Earth in modest amounts. Also, note that the major lunar elements constitute the preponderant mass of mineral elements used in Earth industry (excepting air, water, and fuels).

The selection of chemical or physical processes to convert raw or beneficiated lunar or asteroidal ore (or, alternatively, specialty earth materials) to desired elemental and compound materials suitable for further industrial processing involves many of the same factors of cost, raw material availability, transportation, environmental and personnel hazards, etc., which influence selection and design of Earth-based plants, although the criteria are weighed differently. In addition, the unique constraints and opportunities of the space environment must be considered in selection of suitable

Table 2
RANGES OF CHEMICAL COMPOSITIONS FOR THE MAJOR MINERALS

High-titanium Basalts

Modal abundance (vol %)	Pyroxene 42—60%	Olivine 0—10%	Plagioclase 15—33%	Opaques (mostly ilmenite) 10—34%
Component (wt %)				
SiO_1	44.1—53.8	29.2—38.6	46.9—53.3	< 1.0
Al_2O_3	0.6—6.0	—	28.9—34.5	0—2.0
TiO_2	0.7—6.0	—	—	52.1—74.0
Cr_2O_3	0—0.7	0.1—0.2	—	0.4—2.2
FeO	8.1—45.8	25.4—23.8	0.3—1.4	14.9—45.7
MnO	0—0.7	0.2—0.3	—	< 1.0
MgO	1.7—22.8	33.5—36.5	0—0.3	0.7—8.6
CaO	3.7—20.7	0.2—0.3	14.3—18.6	< 1.0
Na_2O	0—0.2	—	0.7—2.7	—
K_2O	—	—	0—0.4	

Low-Titanium Basalts

Modal abundance (vol %)	42—60%	0—36%	17—33%	1—11%
Component (wt %)				
SiO_2	41.2—54.0	33.5—38.1	44.4—48.2	< 1.0
Al_2O_3	0.6—11.9	—	32.0—35.2	0.1—1.2
TiO_2	0.2—3.0	—	—	50.7—53.9
Cr_2O_3	0—1.5	0.3—0.7	—	0.2—0.8
FeO	13.1—45.5	21.1—47.2	0.4—2.6	44.1—46.8
MnO	0—0.6	0.1—0.4	—	0.3—0.5
MgO	0.3—26.3	18.5—39.2	0.1—1.2	0.1—2.3
CaO	2.0—16.9	0—0.3	16.9—19.2	< 1.0
Na_2O	0—0.1	—	0.4—1.3	—
K_2O	—	—	0—0.3	—

Highlands Rocks

Modal abudance (vol %)	5—35%	0—35%	45—95%	0—5%
Component (wt %)				
SiO_2	51.10—55.4	37.70—39.9	44.00—46.0	0—0.1
Al_2O_3	1.00—2.5	0—0.1	32.00—36.0	0.80—65.0
TiO_2	0.45—1.3	0—0.1	0.02—0.03	0.40—53.0
Cr_2O_3	0.30—0.7	0—0.1	0—0.02	0.40—4.0
FeO	8.20—24.0	13.40—27.3	0.18—0.34	11.60—36.0
MgO	16,70—30.9	33.40—45.5	0—0.18	7.70—20.0
CaO	1.90—16.7	0.20—0.3	19.00—20.0	0—0.6
Na_2O	—	—	0.20—0.6	—
K_2O	—	—	0.03—0.15	—

Modified from Williams, J. R. and Jadwick, J. J., Eds., NASA Reference Publication 1057, Handbook of Lunar Materials, February 1980.

Table 3
ASTEROID SURFACE MATERIALS:
CHARACTERIZATIONS[3]

Asteroid	Mineral assemblage[a]
Hebe	NiFe > Cpx
Iris	NiFe, Ol, Px
Flora	NiFe ≥ Cpx
Metis	NiFe, (Sil [E])
Hygeia	Phy, Opq (C)
Parthenope	NiFe, (Sil [E])
Irene	NiFe, Px
Eunomia	NiFe ~ (Ol ≫ Px)
Psyche	NiFe, Sil (E)
Thetis	NiFe, Cpx
Melpomene	Sil (O), Opq (C)
Fortuna	Phy, Opq (C)
Phocaea	NiFe, Px, Cpx
Euterpe	NiFe, Px, Cpx
Bellona	Sil (O), Poq (C)
Europa[b]	Phy, Opq (C)
Concordia	Phy, Opq (C)
Ausonia	NiFe, Px
Eurynome	NiFe ~ Cpx
Sappho	Sil (O), Poq (C)
Alkmene	Sil (O), Opq (C)
Io[b]	Sil (O), Opq (M)
Thisbe	Sil (O), Opq (C)
Siwa	NiFe, Sil (E)
Dembowska	Ol, (NiFe)

Note: Mathematical symbols ("$>$", greater than; "$\gg$", much greater than; "$\sim$", approximately equal) are used to indicate relative abundance of mineral phases. In cases where abundance is undetermined, order is most abundant to least abundant.

[a] Mineral assemblage of asteroid surface material: NiFe (nickel-iron metal); Ol (olivine); Px (pyroxene, generally low-calcium orthopyroxene); Cpx (clinopyroxene, calcic pyroxene); Sil (O) (mafic silicate, most probably olivine); Sil (E) (spectrally neutral silicate, most probably iron-free pyroxene, enstatite); Phy (phyllosilicate, layer lattice silicate, meteoritic clay mineral, generally hydrated, unleached with abundant subequal Fe^{2+} and Fe^{3+} cations); Opq (C) (opaque phase, most probably magnetite or related opaque oxide).

[b] Satellites of Jupiter.

Modified from Gaffey, M. J. and McCord, T. B., Table III, in Asteroid surface materials: a minerogical characterization from reflectance spectra, *Space Sci. Rev.,* 21, 555, 1978. With permission.

processes. One must establish criteria to compute figures of merit for alternative processes to allow selection of an optimum process route. Once the process route is chosen, the process can be analyzed in terms of the individual unit operations which can then be sized using conventional engineering procedures. The resulting preliminary design can then be compared with other process routes (or with earth-based processing and launch-to-orbit alternatives) to establish systems performance of the various options.

Table 4

POTENTIAL AVAILABILITY OF LUNAR ELEMENTS

Major elements ≥ 1% lunar
O, Si, Al, Ca, Fe, Mg, Ti

Minor elements 0.1—1%
Cr, Mn, Na, K, S, P

Trace elements < 0.1%
H, He, C, N plus all others

Note: Beneficiation may permit concentrating some trace elements into minor or higher range.

B. Criteria for Process Evaluation

The prime consideration for evaluation of space processing and manufacturing systems must center on cost effectiveness in producing structures, functional hardware, and supplies in orbital or lunar surface locations vs. Earth-based processing and launch into orbit. In such comparisons, especially for lunar materials processing, it is essential that functional substitutions be considered since some items such as organics, fiber-reinforced resins, beryllium products, copper, silver, refractory, and precious metals, plus materials with appreciable water content would be difficult to produce from lunar materials. Fortunately, acceptable substitutes exist for any of those substances which would be needed in substantial amounts.

The cost in orbit of Earth manufactured products may be taken as the earth market price plus the cost of orbital lift. The latter is anticipated to be in the range of $654/kg using space shuttle technology.* Several estimates of the cost of launching lunar materials into orbit have been given. It seems safe to conclude that regardless of ultimate technological advances, the cost per unit payload of lunar orbital launch is likely to remain at 10% or less of the cost of earth orbital launch given a sufficient mass requirement.

Regardless of the materials processing and manufacturing systems chosen, it is unrealistic to anticipate that the value added per conversion step of high specific output processes will be comparable or lower in cost than similar operations on the Earth. Exceptions may occur for operations uniquely dependent on the space environment. On the other hand, it is fully reasonable to anticipate that the cost of such operations should not exceed ten times their equivalent cost on Earth. To meet such a limit, it is essential that the mass of capital equipment, expendables, reagent inventory, and support facilities which must be launched from Earth should be far exceeded by the annual output mass of such high rate operations. Analysis to date of the materials processing portion of such operations shows that this requirement is readily met.

The total "Earth supplied equivalent mass" (ESEM) per unit mass output chargeable to the materials processing portion of an industrial facility must be property defined to permit inter-comparison between alternative processing systems. Mass derived from lunar sources should be separately assessed as "lunar supplied equivalent mass" (LSEM) which would include input materials in inventory and in the process loop. It seems preferable to charge the output mass inventory to the manufacturing operation except possibly for output materials used captively for plant operations.

Mass for necessary support services may originate jointly from Earth-supplied and lunar-supplied material. Rather than attempt to break down such items, it may be preferable to assign probable cost figures to such support operations.

---

* Rough costing in 1977 dollars based on payload for low inclination (28°) orbit (NASA Report NSC-12973 Solar Power Satellite Concept Evaluation, July 1977).

Output mass must be clearly defined in terms of products and primary needs. A large mass output of slag-like materials which may only be useful for radiation shielding should perhaps not be listed as a primary output product in computing plant mass requirements per unit "output" mass.

In addition to mass considerations, other criteria of importance in process evaluation include process reliability, manpower requirements for operation and maintenance, potential hazards to on-site personnel, adaptability to process scrap materials, and ease of repair in case of malfunction. In the latter case, corrosion of parts which can only be replaced from earth supply is far more serious than corrosion of lunar derived parts.

Original cost of chemical process equipment per unit mass is expected to be dwarfed by orbital lift costs in all but a few special cases and thus would be of minor importance. If replacement items for many of these units could be fabricated from lunar materials, this would offer the opportunity for cost reduction in growth or replacement in space industrial operations.

C. Process Constraints

A successful orbital or lunar surface materials processing plant must operate with several constraints which rarely concern industrial plants operating on earth. These include

1. Lack of virtually inexhaustible supplies of air and water
2. Lack of unlimited heat sinks offered by (1)
3. Lack of unlimited fuel supplies; coal, oil, electric, gas, etc.
4. Lack of inexhaustible oxidizing and reducing agents
5. Lack of expendable acids and bases (except CaO)
6. Lack of key chemicals: ammonia, salt, chlorine, caustic, soda ash, carbon dioxide, sulfuric and phosphoric acids, carbon and graphite, and organics
7. Lack of ordinary solvents
8. Lack of unlimited inertia in foundations (except on moon)
9. Lack of support vendors

These constraints do not prevent use of these reagents, supplies or services, but make it essential that for high rate processes, ordinarily expendable materials must be recycled to original form with a minimal attrition or loss (preferably below 1% per cycle).

This requirement has a corollary in that the output or material leaving the high mass rate plant must be a separation and/or recombination of the chemical elements present in the feedstock. Since the only nonmetallic elements present in significant quantities from various lunar raw materials are silicon and oxygen, the output streams must be necessarily limited to elements, alloys, silicides, and oxides.

For the major mineral constituents of lunar rock and soil—pyroxenes, feldspars, and olivine—the compositions are silicates which may be described as addition compounds of metal oxides and silica. Conceptually the processing of such materials may be broken down into separation of the constituent oxides (including silica) followed by reduction of that portion of the metallic oxides and silica desired to obtain structural metals and oxygen (or higher oxides, e.g., Fe_2O_3). For ilmenite, $FeTiO_3$, the same steps are necessary except that no silica is involved.

Despite the compositional constraints imposed by lunar source materials, a surprising variety of industrial materials could be supplied. Table 5 shows a list of useful products, with examples of what could be made at a space manufacturing-facility, primarily from lunar materials. Although not listed, water would be made from oxygen obtained from

Table 5
USEFUL PRODUCTS DERIVABLE PRIMARILY FROM LUNAR SOURCES

Structural Materials

Metals—steels, aluminum, magnesium, titanium and alloys
Reinforced metals—metals above reinforced with silica, steel, alumina, or titanium silicide
Glasses—calcium, magnesium, aluminum, or titanium silicates, fused silica, foamed glasses
Ceramics—alumina, magnesia, silica, complex oxides, fused basalts
Hydraulic cements—(need water)

Thermal and Specialty Materials

Refractory and hard materials—ceramics above plus chromia, titania, titanium silicides
Abrasives—alumina, garnets, silicon carbide, titanium carbide (limited by C)
Insulation—ceramics above plus fiberglass, fibrous or powdered ceramics

Electrical Materials

Conductors—aluminum, magnesium, iron, resistance alloys (FeCrAl), silicon
Electrodes—Fe_3O_4, TiO, graphite (limited by C)
Magnetic materials—iron alloys, magnetic ceramics (ferrites, magnetoplumibites)
Electrical insulation—see glasses, ceramics, and thermal insulation

Fibrous Materials

Glass, silica, synthetic mineral wool for apparel, paper, filters, etc.

Plastics and Elastomers

Silicone resins (limited by C)

Sealants, Adhesives, and Coatings

Soluble silicates
Anodized coatings—on aluminum, magnesium, titanium
Electroplating—chromium, etc.
Sputtered or vacuum deposited coatings

Lubricants, Heat Transfer Fluids

Sulfides, graphite (limited by C) SO_2, He

Industrial Chemicals

Detergents, cleansers, solvents, acids, bases H_2SO_4, H_3PO_4, CaO, NaOH

Biosupport

Oxygen (breathing), 16/18 of water by mass SiO_2—soil component (including trace nutrients); Bioelements O, Ca, C, Fe, Mg, K, P, N, Na, H others

lunar materials and hydrogen brought from Earth. (Hydrogen is also a trace element on the moon, but even though its weight abundance may typically range between 50 to 100 ppm, its atomic abundance may be 1% that of silicon. If extractable, this hydrogen could be used to produce the water needed to replace that lost in process recycling, since no recycling process is 100% efficient.)

Table 6
CHEMICAL PLANT DESIGN

Selection of process options
Special space environmental factors
    Gravity (natural or artificial)
    Vacuum
    Heat rejection
    Recycling of nonlunar indigenous materials
Description of unit operations
    Materials handling
    Phase separations
    Heat exchange
    Reactors
    Energy requirements
    Heat rejection requirements
Sizing Factors
    Kinetics limited
    Heat transfer limited
    Momentum limited

The structural metals listed in Table 6 (Al, Fe, Mg, and Ti) will have to be alloyed in order to develop useful mechanical properties. Table 1 shows that several elements (e.g., Si, Cr, Mn) used in forming commercial alloys of the structural metals are potentially recoverable from the moon. In addition, there is "neutral" iron on the moon that also contains nickel and some cobalt, which could possibly be recovered. Finally, if enough carbon were recoverable from the trace amounts present on the moon (or brought from Earth, or possibly obtained from a carbonaceous asteroid), and small amounts of other key alloying elements were imported from Earth, one could produce many ferrous and nonferrous alloys that are commonly used today.

In the case of steels, the more important alloying elements not readily available on the moon are C, Ni, Mo, W, V, and Nb (Cb). The physical properties of several commercial alloys containing these elements are generally similar to alloys in which such elements are absent but which instead contain lunar-indigenous elements. (The properties compared included tensile strength, yield strength, hardness, and elongation.) Therefore, for use as structural metals in space, it is possible to produce alloys possessing a broad range of properties (as commonly required on Earth) by alloying with carbon, lunar-indigenous elements, and/or minor amounts of lunar-deficient elements (LDE).

Several commercial aluminum alloys can be made from lunar-indigenous materials. However, higher strengths, approaching those of the strongest aluminum alloys made on Earth, can only be obtained by alloying with small amounts of lunar-deficient elements, particularly zinc.

Only a few commercial alloys of magnesium and titanium can be formed solely from elements recoverable from the moon; but here again, as in the cases of steel and aluminum, significant improvements in properties can be obtained by alloying with minor amounts of lunar-deficient elements. The strongest magnesium alloys will require such lunar-deficient elements as zinc, while the strongest titanium alloys will need molybdenum.

It is evident that the structural metals to be manufactured in space will consist of alloys that are already in use, are well characterized, and can furnish almost any desired property currently available. Nonmetallic materials similarly may be made solely from lunar sources, or modified with additions of lunar-deficient elements.

III. GENERAL CLASSIFICATION OF MATERIALS PROCESSING SYSTEMS

In an attempt to review and discover practical materials processing systems for lunar or other materials, it seemed worthwhile to attempt a general method of classifying such systems. Although the number of possible process variables is extremely large, especially in composition of one or several reagents, there are certain features in common which distinguish methods of separating constituent components from relatively nonvolatile solid compounds and mixtures. These separation methods and operating temperatures are more important in determining the character and nature of the processing plant than the specific reagents involved.

To separate one or more constituents from a high melting point solid with negligible vapor pressure at ordinary temperatures, one must create conditions to allow the desired constituent to gain a reasonable mobility or diffusibility (M/D) on a micro- or atomic-scale to permit it to react or migrate so that separation may be effected. (In this context, we shall not consider grinding or dispersion techniques which merely serve to reduce particle size or permit transport of solid phases by slurries or fluidization techniques). Figure 2 shows a classification of mobility/diffusibility routes in which the output streams are designated as V (vapor), L (liquid), or F (fluid = V or L). (The intermediate output of the chemical solid-solid reaction is labeled C*, which designates a solid state reaction the solid output of which is rerouted through the system to some other mobilizing step.)

Once a mobile (liquid or vapor) phase is available, a separation from other mobile constituents and residues (which we shall define as residual solid material of negligible vapor pressure) may be accomplished by one of the routes shown on Figure 3. (The residues if present may be recycled to the mobility/diffusibility [M/D] system.) The separation steps are designated by P (physical), S (semi-physical), C (chemical), or EC (electrochemical).

In many cases, the mobilizing and separating operations may take place simultaneously or in the same apparatus. Recycling of chemical reagents may follow the same general outline, although in many cases the volatility or fluidity of the reagents may already be established.

A. Flow Chart Analysis

To further expand the analysis, one may separate an entire chemical or materials processing system into a network (flow chart) of steps or segments, each one of which may be characterized by one or more input streams and one or more output streams. If we denote by (I,O) the number of input and output streams, a (1,1) segment represents either a materials mover such as a pipe, pump, conveyor, etc., or a stream heater, cooler, grinder, crusher, or physical treatment unit. A flow chart is the representation of such a network in which the (1,1) segment representing flow lines are usually drawn as simple lines. Any segment with two or more outputs must of necessity incorporate some phase separation function except for special cases based on differential concentration of a single phase as, for example, with gaseous diffusion units. We may also distinguish between physical segments and chemical (reactor) segments according to whether lack of or presence of a chemical reaction occurs in such steps. Finally, we may note that a mobility/diffusibility step is only required when solids with little or negligible vapor pressure must be treated to permit extraction or flow of a desired constituent and when surface reactivity of the grain is expected to be too slow.

If reasonably pure output products are desired (e.g., 99% purity), it is clear that for processing a typical lunar soil (Table 2); the seven major elements will require at least

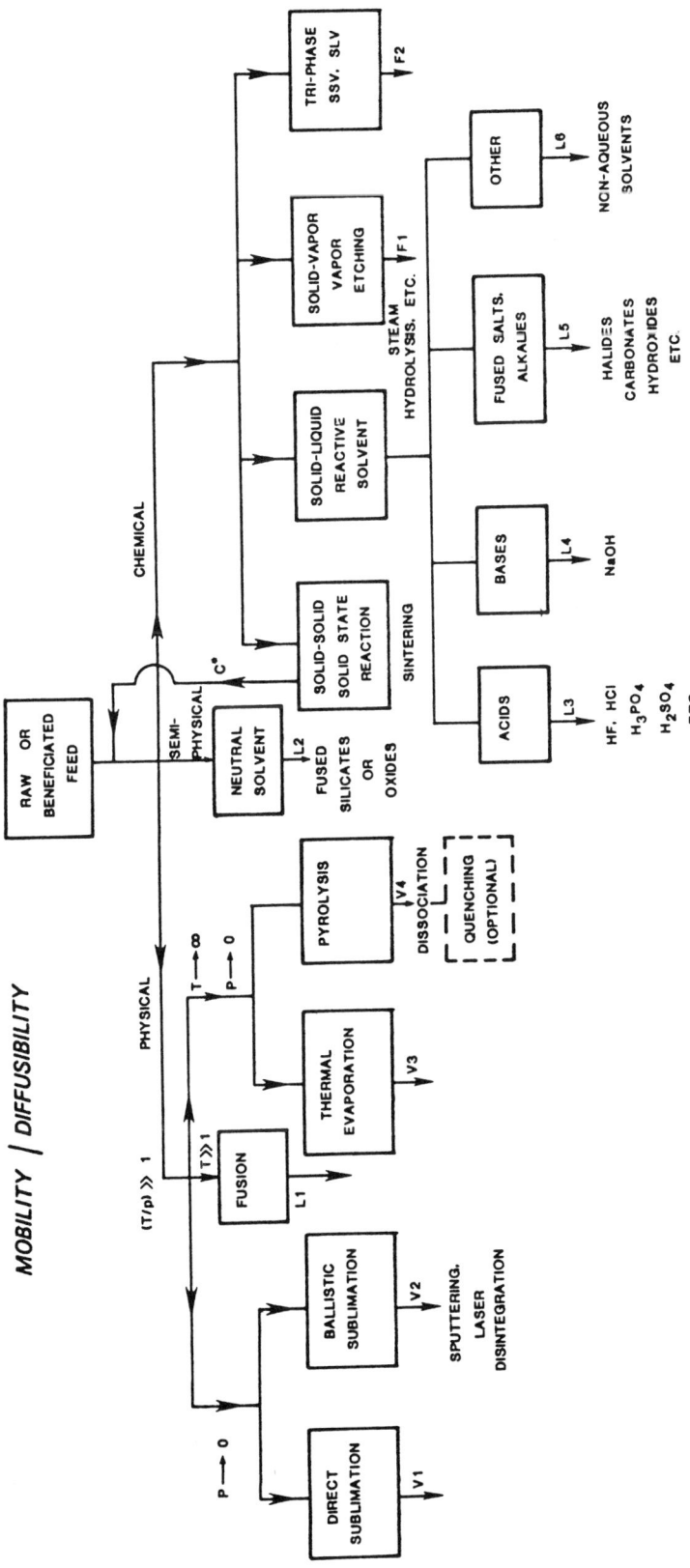

FIGURE 2. Mobility/diffusibility process steps. (From NASA, Lyndon B. Johnson Space Center, Houston.)

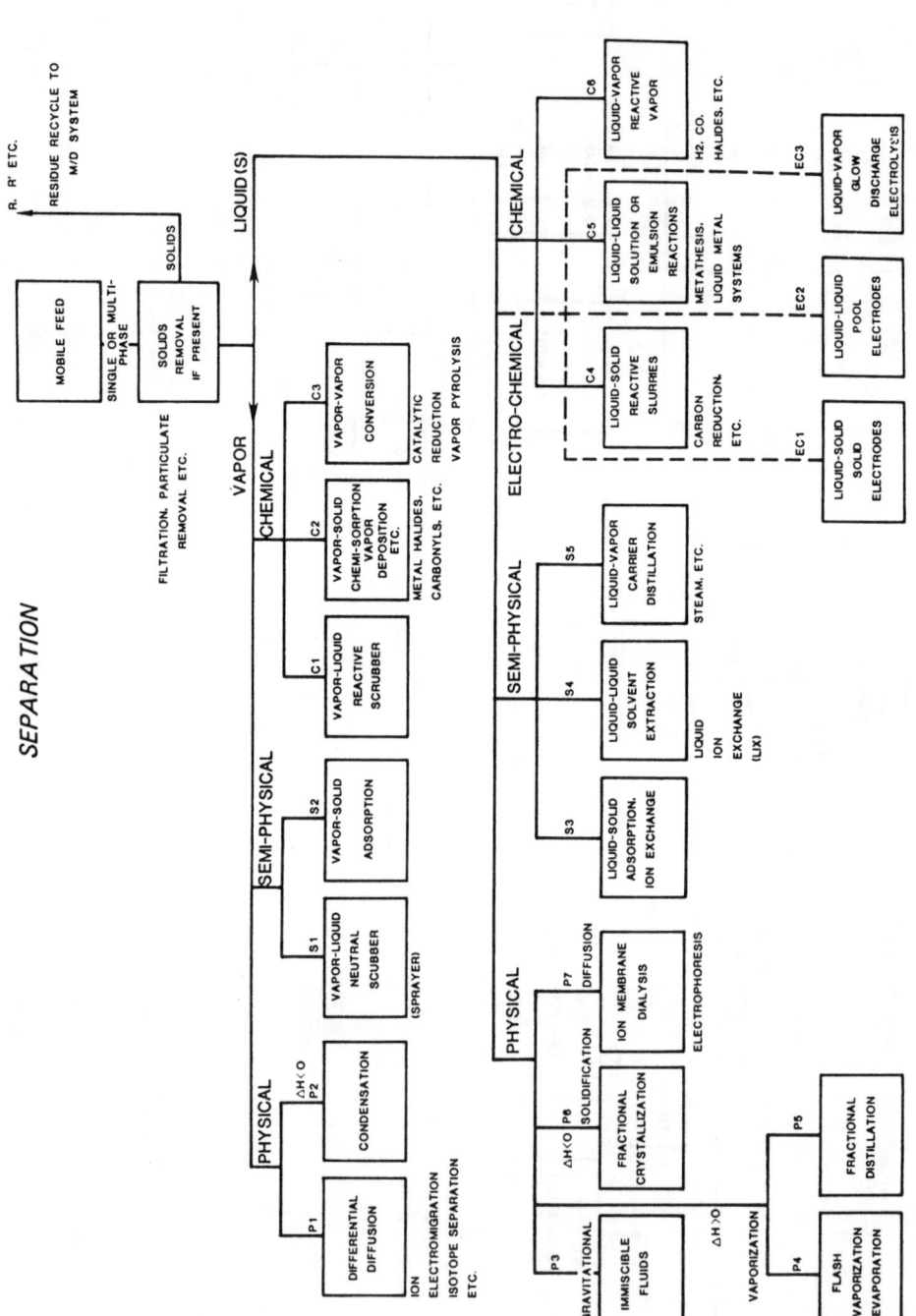

FIGURE 3.   Process fraction separation steps. (From NASA, Lyndon B. Johnson Space Center, Houston.)

six separation steps (assuming [n,2] steps) if all these elements are desired either in reduced or oxide form. Even if one is only interested in recovering oxygen, silicon, aluminum, and iron in commercially pure form, it would appear that at least four separation steps would be required. In addition, extra separation steps may be anticipated for recycling of necessary reagents.

The enormous number of process variations possible may be realized when one considers that even if a single separation process for each of the 22 categories shown in Figure 3 was considered for each of the six separations above, one would have more than 113 million combinations (22^6) of separation segments to consider. The 13 classes of M/D steps would further increase the number of possible process variations.

The complete network or flow diagram must contain steps necessary to recycle all reagents not derivable from lunar soils. A detailed mass balance chart should also include mass replacements for electrode attrition, wear or corrosion of containers, etc. These latter considerations in many cases may require a greater mass replacement rate than reagent recycle loops with readily achievable efficiencies.

In order to narrow the field of promising materials processing systems and limit the cost and scope of development effort expended in analysis and improvement of parallel processes, it is imperative that a method of rating comparable processes be established based on realistic evaluation of anticipated performance and realistic assessment of technological risks involved.

B. General Survey or Overview of Processing Methods

For the M/D sections, the physical options have an advantage over the other routes in not requiring reagents or solvents. On the other hand, either very high temperatures and/or low pressures or high energy excitation is necessary to vaporize or fluidize silicate rock. Vapor pressures below about 0.1 Torr so restrict materials transfer rates that they are seldom of process interest for high volume production. In systems at very high temperatures, thermal losses at low pressures are apt to be excessive relative to mass transfer rates.

Neutral* solvent systems (L2) would normally be employed rather than fusion (L1) if a substantial reduction in melting point or operating temperature is possible. There would be little incentive to use a neutral solvent at operating temperatures near the melting point of the feed material. Reactive** solvents (L3-6) can operate from at or below room temperature up to high temperatures, but the latter would probably not be useful for the same reasons as with neutral solvents. All of the chemical systems listed here are considered to operate below the melting point of the feed material. For higher temperature chemical operations, the feedstock is presumed to have melted (L1) and the chemical reaction is treated in the separation section.

For the separation section, the physical and semiphysical options involve well-recognized phase and homogeneous separations. The diffusion routes (P1 and P7) are often not too highly selective at useful concentrations and are generally employed only when no alternative phase separation is practical or convenient. The two phase scrubbers and absorbers (S_1, S_2, C_2, S_3) are often useful at low to intermediate temperatures but are apt to present severe materials problems at elevated temperatures.

C. General Observations Concerning Chemical Conversions

The previous classification does not separately group steps to produce free elements

---

*A neutral solvent is defined as one from which the original material may be recrystallized (in principle) in a substantially equivalent form to its original state.

**Reactive solvents produce chemical alteration of one or more constituents of the solute.

or other reduction products. These do not differ materially from other chemical steps and require the same types of separation procedures. Metals reductions at temperatures above their melting points are generally self-separating due to the common immiscibility of molten metals with nonmetals, slags, or fused salts. (This is not true in all cases, however.) However, the separation of constituent elements of alloys is generally difficult.

Solvent systems may be subdivided into aqueous (L_3, L_4) and nonaqueous classes. In the former, the water solubility in acid and near neutral solutions of metallic compounds is of general interest. Note that most metallic nitrates, chlorides, perchlorates, fluoborates, and fluosilicates are water soluble, while there is some limited solubility of fluorides and sulfates for some of the metals. Most of the remaining common anions form insoluble salts with all but the alkali metals. Nitrate, perchlorate, and fluoborate compounds present stability or availability problems that render them less attractive than the other salts for general separation and reduction operations.

In basic solution, only the silica, titania, and alumina fractions are likely to have sufficient solubility to be of interest, and these constituents are often rendered insoluble in the presence of polyvalent metallic ions such as Ca^{++}, Mg^{++}, etc. Recycling of base (NaOH) is somewhat more difficult than for acids due to lower volatility.

Most metal chlorides are readily vaporized, while fluorides are much less volatile. Both are easily reduced by active metals or electrochemical action. Sulfides are also potential candidates for metals reduction, but many have very high melting points.

D. Reduction Processes

Reduction processes may be expected to differ in significant aspects depending on the element being recovered. The principal routes may be divided into direct (electrochemical reduction) methods and indirect (pyrochemical) methods; however, the latter would generally require electrochemical regeneration of the reducing agent. For processing on the lunar surface, it may be practical to use neutral iron as a reductant with ferrous or ferric oxide as spent product without attempting to recycle the iron. Such open cycle use of reductants, which is commonplace on earth, would be prohibitive in orbit due to high launch costs even from the moon.

Direct electrodeposition processes may be conducted in a variety of electrolytes, but only a small fraction of lunar metallic content may be deposited from aqueous solution. Of the major and minor lunar elements, only Fe, Mn, and Cr are normally platable from aqueous solutions,[4] although Na, K, Ca, and Mg could probably be recovered by using a mercury cathode. Aluminum might be recovered with ternary liquid metal alloy cathodes, but its solubility in mercury appears to be too low for practical operation.

All of the metals are recoverable by electrodeposition from various nonaqueous electrolytes, principally fused salts, but these processes pose a number of corrosion and anode durability problems, depending chiefly on the temperature of operation. Electrodeposition from electrolytes containing two or more of the reducible elements may also present formidable purification problems.

It would be highly desirable to generate oxygen at the anode instead of chlorine or other product, since reconversion of the chlorine to chlorides and evolution of oxygen in some recycle step would involve another oxidation-reduction reaction. Electrolysis of fused silicates, carbonates, hydroxides, or oxides, of such compounds dissolved in molten fluorides can generate oxygen, but for such processes conducted above 400 to 500°C, the resistance of potential anode materials deteriorates rapidly. For example, in commercial aluminum production, the use of graphite anodes results in virtually complete conversion of the oxygen to carbon monoxide and dioxide at temperatures between 950 and 1000°C.[5] Despite considerable work in this field, no satisfactory durable anode has been developed for this application.

One may, of course, recycle the oxides of carbon to oxygen and graphite, but this is not an easy operation, and the fabrication of graphite electrodes is a very slow and mass intensive process and should be avoided if at all possible.

Fortunately, satisfactory electrodes with very long service lives have been developed for oxygen evolution from aqueous solutions and from fused alkali hydroxides operating near 300°C. Electrolysis in this latter system was pioneered by Hamilton Y. Castner who developed a process for sodium production nearly a century ago.[6] The Castner cell was subsequently superseded for sodium production, and aluminum production from sodium was discontinued, but a modification of this method appears to offer many advantages for an extraterrestrial reduction process.

This process would generate the required number of reduction equivalents of sodium plus oxygen, and the sodium would be used for indirect (pyrochemical reduction) of silicon and the structural metals. Reduction of magnesium halides with sodium would not proceed to completion under normal circumstances, but magnesium oxide may be reduced with silicon which can be formed by sodium reduction.

E. Summary of Chemical Conversions

The difficulties attendant upon separation and/or direct reduction of constituents of complex oxides and silicates prompts one to examine various classes of compounds which can be generated by treatment of the oxide materials by various reagents. From the previous discussions on aqueous solvent systems, one might wish to consider nitrates, chlorides, perchlorates, fluoborates, fluosilicates, fluorides, and sulfates. Sulfides, carbonates, phosphates, and (to a limited extent) carbonyls might also be usefully employed in certain phases of materials processing loops. Nitrates and perchlorates present potential difficulties due to their instabilities toward severe thermal or oxidation-reduction conditions and to the difficulties in resynthesizing such reagents. Fluoborates seem to offer few advantages in comparison with fluosilicates and require makeup of another lunar deficient element. Sulfates, carbonates, and phosphates seem to present limited capabilities in general solubility/separation operations but might be useful in specific separations.

Ammonia/ammonium salt chemistry has a unique advantage in that pyrolysis of ammonium compounds can usefully purify a number of the major and minor lunar elements as readily convertible compounds. The stability of ammonium ion or ammonia is not as great as halides toward severe oxidation or thermal exposure, but it is better than nitrates or perchlorates and regeneration is fairly easy in comparison with these compounds.

F. Sizing

In the absence of reliable kinetic data, it is difficult to estimate the size and mass of process equipment which will be required to obtain unit output from any proposed process. However, it is possible to list certain features that should probably be avoided or minimized if possible for high specific output plants. These include

1. Steps that require long completion times
2. Steps in which the input material is present in low concentration
3. Mass transport of volatiles at very low pressures
4. Phase separations from viscous suspensions
5. Reactions with low percent conversion per pass
6. Reactions involving handling or storage of large volumes of gas
7. Reactions involving large transfers of heat to or from single phase fluids, especially gases, using heat exchangers

8. Processes which reject large amounts of process heat at low temperatures (below 200 to 300°C)
9. Processes for which suitable structural materials do not offer reasonable service lives

IV. CHEMICAL PLANT DESIGN

Some factors that must be considered in designing a chemical (Materials Processing/Refining) plant in space are shown in Table 6. The specific problems or opportunities created by these factors are expected to differ materially for high specific output plants (for conversion of Lunar/Space raw materials) than for specialty plants (for conversion of Earth materials which will usually be partially refined.

A. High Specific Output Plants for Conversion of Space-Derived Materials
1. General Considerations
A chemical plant for extraterrestrial materials processing may be expected to utilize equipment very similar to that employed in earth-based plants. Because of the importance of minimal mass, most apparatus initially brought from earth will be constructed of materials of high specific strength (strength/weight ratio) perhaps using thin linings of corrosion resistant materials (e.g., even gold). Later equipment made from lunar materials would not require extraordinary strength/weight ratio materials. Special consideration may also be required to be compatible with the special space environmental factors encountered during transport, asssembly, and operation. These include unlimited vacuum sink, adjustable level or artificial gravity (except on lunar surface), and provision for radiative dissipation of process heat loads.

2. Space Environmental Factors
a. Vacuum
The vacuum sink availability for space processing facilities may be useful for several types of operations either in orbit or on the lunar surface. The most generally useful would be the ability to use refractories and structural materials which are normally sensitive to oxidation at higher temperatures than would otherwise be possible except inside vacuum furnaces. Thus, ordinary steels could be used for retorts in metals reductions, and such materials as titanium and refractory metals, carbon and carbides, boron nitride and other nonoxide refractories could be used for structural and insulation purposes without danger of excessive oxidation. This should permit improved multilayer radiation shield insulations for extremely high temperature processes. Sublimational effects may limit the utility of such systems for certain applications, however.

The use of space vacuum as a separation technique may have very limited application, since the escape of volatiles except for very limited amounts of oxygen or water vapor could rarely be tolerated.

It would appear desirable to locate most of the processing facilities in a large container with an atmosphere and temperature compatible with human activity. This would permit easier inspection, maintenance and operation of the system and thus greater productivity.

b. Gravity
The reduced gravitational attraction for lunar-based plants or adjustable centrifugal forces for simulated gravity in orbital plants will allow some mass savings in support structures for process equipment. It seems likely that most of the chemical unit operations would not operate satisfactorily under conditions of weightlessness, since all

mass transfer operations except for introduction of gases into a vessel would be unnecessarily complicated by absence of a gravitational effect. Storage tanks or reactors of fixed volume for solids, liquids, or slurries would be difficult to load or unload and such operations as filtration, distillation, countercurrent extraction, etc., would be rendered difficult if not impossible.

The most likely uses for weightless processing would be for heating corrosive reaction masses by radiation or induction using gas jet or electromagnetic repulsion to prevent contact with the walls of a chamber and, after removing volatile products, byproducts or impurities if present, allowing the reaction mass to cool in place or in a "drift tube" zone until it could be handled.

c. Heat Sink

The unavailability of massive external air or water heat sinks makes management of process waste heat especially important. All major heat rejection loads will ultimately have to be transferred to space radiators for final disposition. In addition, the poor heat transfer characteristics of vapor heat exchange devices makes such elements heavy and undesirable. This leads to the general conclusion that to raise or lower the temperature of a gas stream it will be preferable to adiabatically compress or expand the stream rather than use wall or tube type heat exchangers. Similarly, in distillation operations it will be advantageous to use a boiler heated by adiabatic compression. The mass penalty for additional pumping power will usually be far lower than other alternatives for disposition or transfer or process heat.*

Unavoidable low or medium temperature heat loads, such as from electrolytic cells may require heat pumping to higher temperatures to avoid excessive space radiator masses. A simplified analysis indicates that below some temperature determined by mass power ratios of space power systems and mass:area ratios of space radiators, it becomes desirable to heat pump all heat rejection loads to such base line temperatures which fall in the range 500 to 600K based on satellite hardware masses of current design. Similarly, refrigeration equipment for liquefaction of cryogenics should have heat rejection temperatures at the same level.

3. Reagent and Equipment Mass

For solution processes, the mass of the solvent system will generally exceed the mass of lunar input material except where solutions of over 50% by weight are practical. A more typical level may be about 5 to 10% by weight of solute. Futhermore, not all of the solute may be transferred per pass when the various separation or extraction steps are performed, so the ratio of solvent to "active solute" mass is normally much greater than unity.

Fortunately, for aqueous solutions, most of the solvent mass need not be transported from the Earth, since the oxygen content which represents 88.8% of the mass of water, is derivable from lunar materials. Even the hydrogen content may be extracted in sufficient quantities to largely or entirely replace the content lost in residual moisture content of plant products.

One may inquire as to the relative magnitude of equipment and reagent mass for the various units needed for a chemical processing plant. Specifically, one would like to know whether the vessels, tanks, pipes, and other items of process equipment weigh more or less than their contents. A simple analysis shows that for most cases of equipment which contain 10% of more material in condensed phases, the contents may be expected to far outweigh the container, while for gases the container will invariably

---

* See, e.g., Chapman, A. J., Heat Transfer, MacMillan, New York, 1960.

outweigh the contents and furthermore, in this case, the ratio of container to content mass is practically independent of pressure.

This finding reiterates the undesirability of processes which require storage or handling of large volumes of gas. In addition, for processes operating primarily in condensed phases, the mass of the processing operation, apart from power and radiator facilities, will probably be dominated by the masses of reagents involved, which in turn will depend on reaction and process times for the individual steps. In the analyses which follow, we shall estimate both total reagent masses and net (Earth based) reagent masses for some of the process steps of the HF acid leach process.

It shall be convenient in the subsequent analyses to define three mass ratio terms; R_m, R'_m and r_m which are respectively the ratios of the mass of vessel contents, net mass (LDE)* of vessel contents, and mass of container to the equivalent input mass of lunar ore contained in the respective vessel or process apparatus. These ratios multiplied by actual process times yield equivalent process times which will be used in a later section to estimate plant mass and volume (sizing).

4. Unit Operations

The unit operations required to perform the processing steps required for conversion of raw materials into industrial feedstocks are those generally familiar to the chemical engineering profession. These may be grouped into the following classes:

1. Materials handling: storage, conveying, pumping, compression, mixing, stirring, extruding, grinding, metering, etc.
2. Phase separation: distillation, filtration, extraction, drying, defoaming, precipitation, crystallization, sedimentation, centrifugation, etc.
3. Energy, heating and refrigeration: generation of process power and transfer of heat into or out of reactors and other processors
4. Reactors: solid-solid, solid-liquid, solid-gas, liquid-liquid, liquid-gas, and tri or polyphase systems

a. Materials Handling

Except for storage, these operations are not expected to require substantial masses. Most material can pass through a materials handling step with velocities of .01 to 1 m/sec or even higher so such units would rarely have to handle more than a few minutes throughput of the operation. Fine grinding using a ball mill or equivalent may be somewhat slower, but is not expected to be necessary for processing of lunar soils. Entrainment of liquids or dust in gas flows may become a problem in lunar gravity or low artificial gravity, but inertial gas- or hydro-cyclones or other devices may be used to suppress carryover.

b. Phase Separation

The actual physical separation of different phases is usually limited by pressure or inertial considerations such as foaming or entrainment in distillation columns or sedimentation velocities in centrifugal filters or sedimentation centrifuges, although the material process time may be limited by heat transfer rates, growth rates of crystallites or precipitates, etc. Because of the recycle nature of the various materials flow loops, it may be more desirable to shorten process times even at the expense of recycling larger than normal fractions of intermediate flow streams to reduce masses of intermediate stages. However, at the exit stages of the plant, it is important to limit loss of reagents,

---

* Lunar Deficient Elements are all elements except the 13 major and minor lunar elements.

especially those containing LDE, so it is necessary to attempt to carry those steps nearly to completion.

As an example we may consider drying of non-metallic output streams. For many cases in which industrial drying of solids is practiced, the observed drying rate or rate of weight loss is initially nearly constant, but below a certain moisture content, the rate drops and often becomes nearly proportional to the "excess residual water content" or content in excess of the equilibrium level corresponding to the local temperature and pressure.[7] For such a dependence, drying would continue at a progressively slower rate for an infinitely long time and never reach constant weight. For a practical process, the operation must be terminated at some reasonable time or residual moisture content. It may be readily shown that an optimum drying time or moisture content can be evaluated in terms of the minimized total mass of drying equipment and replacement mass to supply hydrogen for water lost.

c. Process Energy, Heating, and Refrigeration

Process energy requirements may be satisfied by primary electrical or solar thermal sources, or indirectly using steam or other working fluid or by exchange with other process flow streams. For processes in which solar thermal energy is possible, one may anticipate a substantial mass reduction for equivalent power levels. (Mass reductions by factors of 30 or more may be possible by substituting solar thermal for solar electric power.)

The coupling of thermal energy into powdered solids is often a troublesome task, and can rarely be done efficiently by radiation. One would normally prefer to heat such material by exchange with recirculating gases heated in an adjacent unit by contact with structures heated by a solar furnace, electric arc, resistance, or induction sources. In certain cases, it may be possible to heat the powdered solids by high-frequency dielectric or microwave energy.

In heat exchange in which gas flow in one or both streams plays a part, one would like to operate at very high velocities or Reynold's numbers since the heat transfer coefficients in turbulent flow are roughly proportional to the 0.8 power of velocity or Reynold's number.[8] Heat transfer involving fluids in boiling or condensing flow are much higher than when no phase change is involved, so when liquids must be heated it is advantageous to operate under conditions of solution pressure and heater temperature to produce nucleate boiling at the interface when vapor pressures permit such operating modes.

Refrigeration or cooling operations may be required for process steps or for collecting, separating, and storing noncondensible gases. Oxygen storage and hydrogen storage will probably represent the largest power and equipment requirements. Liquefaction of these gases would greatly reduce masses of the storage vessels required to handle these materials. By subcooling down to the triple point or lower, even further weight reductions are possible.

d. Reactors

The design of reactors is usually dictated by the heat balance requirements (endothermic or exothermic) and whether internal or external heating or cooling are required. Internally heated or cooled systems can usually be designed in large tubular, cylindrical, or spherical vessels, while external heat transfer usually requires a large surface area and at least one short dimension (about 0.2 to 0.5 m). Electrolytic cells usually require a low anode-cathode separation (about 0.1 m or less) to avoid excessive power losses, but the cell may contain multiple anodes and cathodes and thus attain considerable minimum dimensions. Heat rejection requirements usually limit the size of electrolytic cells, however.

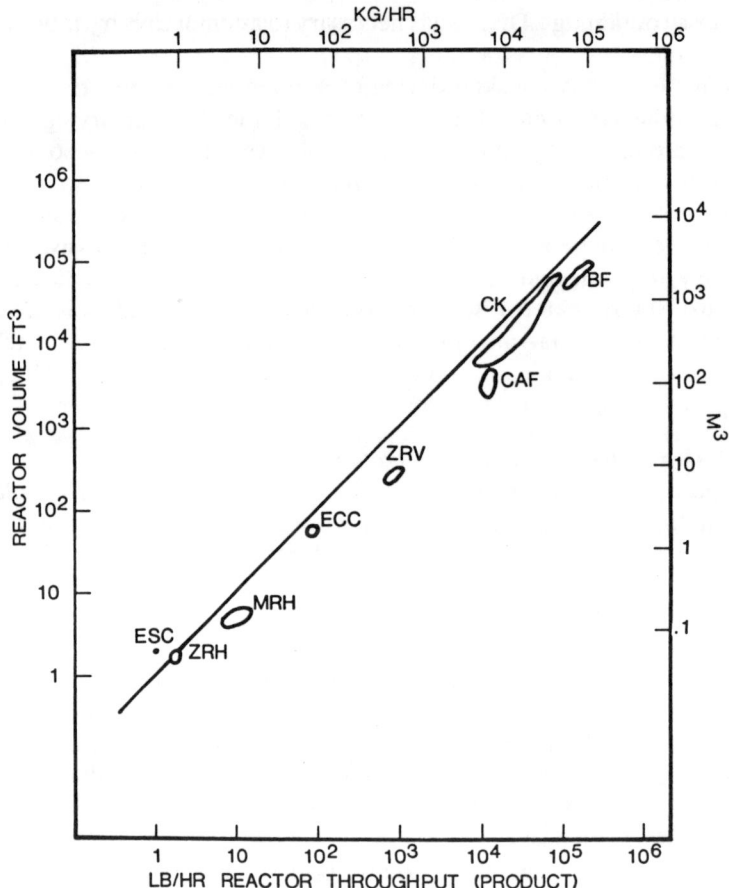

FIGURE 4.   Reactor sizing: BF—Blast Furnace; CAF—Carbide Arc Furnace; CK—Cement Kiln; ECC—Electrolytic Chlorine Cell; ESC—Electrolytic Sodium Cell; MRH—Magnesium Retort (Horizontal); ZRH—Zinc Retort (Horizontal); ZRV—Zinc Retort (vertical). (From NASA, Lyndon B. Johnson Space Center, Houston.)

It is somewhat remarkable that the mass output of a diverse range of chemical reactors per unit volume per unit time in sizes that span over five orders of magnitude are nearly constant lying close to 1 lb/ft³ hr in English units or about 16kg/m³ hr. Figure 4 shows a graph of several reactors from blast furnaces and cement kilns at the large end to electrolytic cells and magnesium and zinc retorts with volumes below .05 m³.* This chart can be used to predict or verify the size of reactors for lunar materials processing steps estimated in the absence of commercial or pilot plant qualifying data.

Corrosion attack on reactor structures or general thermophysical deterioration may be expected to some degree in all high temperature processes, except where cold wall systems are used. This reality may present the greatest deterrent to use of very high temperatures in materials processing steps. The use of valves, pumps, filters, materials handling, and other equipment and the containment of pressures becomes exceedingly difficult at temperatures above 1500°C. Reliability of operations may be adversely affected, and maintenance requirements excessive if many operations are carried out at

* Data compiled from Reference 15 and other sources.

such temperatures. In contrast, suitable apparatus and materials have been developed for handling almost any substance present in water solutions or steam-based systems.

5. General Sizing Considerations

The size of chemical reactors and other process equipment is generally controlled by one or more of three factors: reaction kinetics; heat transfer limitations (surface area); or momentum limitations in which the inertial effects of mass movement may cause foaming, entrainment, or turbid dispersion of multiphase systems. The reaction rates in heterogeneous systems are often diffusion limited, but the use of fine particles and high turbulence can increase throughputs in gas-liquid and gas-solid systems. For reactions involving crystal or precipitate growth, only the degree of supersaturation or control of nucleation can markedly affect the process rate. The rates of most chemical reactions can be increased by raising the temperature, but the equilibrium constant or conversion fraction may be adversely affected for some cases. This may also require higher pressure apparatus which will then require more massive reaction vessels.

The engineering characterization of any proposed process may be identified by parameters defined below. The sizing (volume) of equipment for any segment may be expressed as:

$$V = \frac{\dot{Q}_i \ell}{\rho_i \, v_i} = \frac{\dot{Q}}{\rho_i} \; t$$

where $\dot{Q}_i$ is the mass flow rate of component i (kg/sec), ℓ is a characteristic length (meter) of flow path in apparatus, ρ_i is the partial density of component i (kg/m³), v_i is a characteristic velocity of the ith component (m/sec), and t is the process time (seconds). Parameter values characteristic of commercial process equipment may be derived from apparatus specifications. This will permit sizing estimates for specific processes not presently in commercial or pilot plant service.

The equipment and reagent masses corresponding to the process volumes can be derived using the mass ratio terms and equivalent times previously defined. The equivalent times represent the times required for passage of equivalent lunar input material (mass) equal to the mass of the specific vessel, equipment, or reagent contents for the process step in question. Summation of the appropriate equivalent times therefore yields the total time required for the passage of sufficient input raw-material to equal the gross or net reagent masses or the structural masses for the processing system. Additional equivalent times may be derived to account for masses required for mechanical and thermal power sources; for distribution equipment, motors, pumps, compressors; for heat-transfer equipment, space radiator facilities; and for other necessary support functions.

B. Specialty Plants for Conversion of Earth-Derived Materials

As previously outlined, materials processing of Earth-derived materials will probably be restricted to materials used in fabrication of high specific value products, and then probably only where the unique properties of the space environment permit substantial improvements in efficiency or capability in comparison with processing on the surface of the earth. (Scrap in space reprocessing systems, regardless of the source of the scrap have been excluded from this category as previously noted.)

The design factors previously listed apply also to the specialty plants, but many of the comments regarding practical limitations need no longer apply. Thus while the special space environmental factors are more apt to be obstacles rather than advantages for high specific output plants, they must provide some unique opportunity to prepare

Table 7
OPPORTUNITIES AFFORDED BY SPACE ENVIRONMENTAL FACTORS

Vacuum
    Processes requiring large structures in vacuum
    Processes requiring multiple vacuum to atmosphere transitions
    Processes requiring slow degassing or desorption (heat sensitive drying, freeze drying) and thus large
        volume vacuum storage or inventory
    Processes requiring extensive electron beam processing
    Processes requiring air sensitive refractories
Low or Zero Gravity
    Containerless processing
        Metals
        Semiconductors
        Insulators
    Processes favored by suppression of convection
        VHT (very high temperature) processes with inert gas fill
        Controlled solidification
            Single phase, single crystal or polycrystalline materials
            Directional solidification
            Zone refining
            Multiphase solidification (alloys with unique dispersions)
    Processes for preparing materials with controlled gradient properties (mechanical, thermal,
        electromagnetic) etc.
Low temperatures
    Processes requiring long duration cryogenic cooling and thus large volume cryogenic storage or
        inventory

materials with purities or compositions either impossible to achieve in an Earth-based plant or whose cost of production on Earth would exceed the space processing cost by at least the transportation penalty.

Table 7 lists some possible classes of operations which could be performed more readily in the space environment.

The best prospects for specialty processing would involve materials of high specific value such as precious metals, gemstones, rare isotopes, and some very high temperature materials. In addition, some organic compounds of biological interest such as enzymes, hormones, or compounds of pharmaceutical interest may be candidates.

Some of the materials processes which have been proposed for space facilities have included: growth and purification of semiconductor crystals, growth of metallic and gemstone crystals, preparation of ultrapure glasses for optical fiber waveguides, preparation of high performance rare earth-cobalt magnetic materials, and preparation of biochemicals by electrophoresis.[9]

1. Containerless Processing

Levitation or confinement of hot matter without a physical boundary has generally been restricted to small masses for research purposes on Earth,[10] although confinement of plasmas has been the subject of intense development for fusion power systems.[11] It is relatively simple to levitate small masses of electrically conducting fluids (molten metals) against Earth gravity, and methods have been proposed which would allow levitation of any desired mass.[12] The suspension is normally achieved by the reaction forces between induced electrical currents in the fluid and external currents. This produces a heating effect in addition to the lifting force, the ratio of which can be adjusted to some extent by varying the frequency or geometry of the alternating field, but it can not be made arbitrarily small. In residual force fields below about 10^{-3} earth gravities, the heating effect can be reduced to almost any desired value without danger of the fluid contacting external apparatus. Thus one may perform various cooling cycles which would be impossible for fluids on Earth. In addition, other methods of suspension using weaker forces can be considered such as electrostatic suspension, acoustic standing waves, photon beams, gas or vapor stream momentum, and the magnetic induction method can be extended to poorly conducting fluids such as semiconductors, molten glasses and refractories, etc. (Most refractory insulators have high enough loss currents in the molten phase to permit suspension in orbital apparatus.)

V. SPECIFIC PROCESSES FOR LUNAR-DERIVED MATERIALS

A. Electrolysis of Molten Silicates

Limited investigations of direct electrolysis of molten silicates of compositions similar to lunar basalts have been performed.[13] The high melting points and viscosities of molten silicates have created problems and prompted studies of various fluxing additions to the melt. This modification, of course, negates the "reagentless" advantage of the direct electrolysis route and requires consideration of extraction and recycling of fluxing reagents.

The chief objections or problems awaiting solution are the corrosion or durability of anodes used for oxygen recovery and the purification and separation of cathodic reduction products which are likely to consist of iron-aluminum-silicon alloys plus minor amounts of additional impurities.

B. Carbothermic/Silicothermic Reduction Process

One of the first serious attempts to define a process option was performed by Phinney et al.[14] at the 1976 NASA-Ames Summer Study, in which silicothermic and carbothermic reduction of bulk lunar soil was discussed. After crushing the raw material, and magnetically separating the ferrous from the nonferrous fractions, reduction could commence. Silicon will reduce iron at 1300°C, as shown by Equation 1:

$$2FeO + Si \rightarrow 2Fe + SiO_2 \tag{1}$$

The products can be separated by centrifugation.

The iron-free silicates would be reduced by carbon at 2300°C, as shown by Equations 2 to 5:

$$MgO + C \rightarrow Mg + CO \tag{2}$$

$$SiO_2 + 2C \rightarrow Si + 2CO \tag{3}$$

$$Al_2O_3 + 3C \rightarrow 2Al + 3CO \tag{4}$$

$$CaO + C \rightarrow Ca + CO \tag{5}$$

By this process, it would be expected that aluminum and silicon form a melt, while the other reduced metals, including the major impurities Ti, Mn, and Cr, would be removed as vapors. However, the reaction chemistry is much more complicated. At 2300°C, condensed compounds, such as SiC, Al_4C_3, and Al_4O_4C are present, along with gases such as Al_2O, SiO, Al, and Si. The equilibrium pressures of Al_2O and Al are so high, that liquid aluminum cannot be formed at pressures near one atmosphere. Perhaps the greatest defect of the carbothermic reduction process is that although the winning of aluminum on Earth via carbothermic reduction has been attempted for many years, no practical process for producing purified aluminum by this method has proven satisfactory.[15]

Table 8
CARBO-CHLORINATION PROCESS EQUATIONS

$CaO \cdot Al_2O_3 \cdot (2.25\ SiO_2) \cdot (0.15\ MO) + 8.65\ C + 8.65\ Cl_2 =$
(1) $CaCl_2 + 2\ AlCl_3 + 2.25\ SiCl_4 + 0.15\ MCl_2 + 8.65\ CO$
(2) $CaCl_2 + 2H_2O + (\pm)\ Ca\ (OH)_2 + H_2 + Cl_2$
(3) $0.15\ MCl_2 + 0.3\ H_2O + (\pm) = 0.15\ M(OH)_2 + 0.15\ H_2 + 0.15\ Cl_2$
(4) $2\ AlCl_3 + (\pm)\ (fused\ salt) = 2\ Al + 3\ Cl_2$
(5) $2.25\ y\ SiCl_4 + 4.5\ yH_2 = 2.25\ y\ Si + 9y\ HCl$
(6) $2.25\ (1-y)\ SiCl_4 + 9(1-y)\ H_2O = 2.25\ (1-y)\ Si(OH)_4 + 9(1-y)\ HCl$
(7) $9\ HCl + (\pm) = 4.5H_2 + 4.5\ Cl_2$
(8) $8.65\ CO + nH_2 = (intermediates) = 8.65C + (n-8.65)H_2 + 8.65\ H_2O$
(9) $Ca(OH)_2 = CaO + H_2O$
(10) $0.15M\ (OH)_2 = 0.15\ MO + 0.15\ H_2O$
(11) $2.25\ (1-y)\ Si(OH)_4 = 2.25\ (1-y)\ SiO_2 + 4.5\ (1-y)H_2O$
(12) $(3 + 4.5\ y)\ H_2O + (\pm) = (3 + 4.5\ y)H_2 + (1.5 + 2.25\ y)O_2$

C. Carbo-Chlorination Process

At the 1977 NASA-Ames Summer Study, Rao et al.[16] decided quite early that carbothermic reduction would probably be impractical for space processing. They opted for carbochlorination of lunar anorthite, $CaAl_2Si_2O_8$, and lunar ilmenite, $FeTiO_3$, which could be beneficiated from lunar soil.[17] The desired products are aluminum, iron, silicon (or silica), and titanium. The basic reactions involved in this process are shown in Table 8. Reaction 1 represents the carbo-chlorination of anorthite, run at 700°C. Electrolysis of $AlCl_3$ would yield aluminum, while hydrolysis of $SiCl_4$ would give silica, or reduction of it would produce silicon. Reactions 13 to 17 show the basic scheme involved in carbochlorination of ilemnite. At 800°C the iron in ilmenite is selectively chlorinated[18] as was shown by Reaction 13. Rao and co-workers felt that a fluidized bed process would achieve the best results.

They also discussed methods for obtaining magnesium and oxygen from nonterrestrial materials. They would extract magnesium from olivine (forsterite, Mg_2SiO_4) via silicothermic reduction at 1200°C, as shown by Reaction 18. Reactions 8, 19, and 20 describe reactions on the effluent CO from carbochlorination, so that conversions can be performed on the resultant CO, CO_2, and H_2O products, in order to produce oxygen. A basic flow chart for the process including carbon and oxygen recovery is shown in Figures 5A and B.

Subsequent studies have indicated that carbochlorination would create a major problem of plant size. The recycling of chlorine and carbon would require facilities much larger than the basic processing plant. One of the major advantages of carbochlorination was that it would require only a minimum of hydrometallurgical opeations. Water would be needed, however, for hydrolysis, chlorine regeneration and possibly a coolant for the system. In order to minimize the size of the heat rejection radiators, large amounts of heat energy would have to be raised (heat pumped) to about 280°C, to achieve a mass efficient system.

These results led us to believe that processes which rejected less heat at low temperatures and making use of hydrometallurgical operations would tend to be more useful options for space processing.

D. NaOH Basic Leach Process

A review of the literature[19-22] indicates that anorthite can be decomposed with NaOH in an autoclave, and subsequent treatment of those products with more base can eventually yield alumina and calcium silicate; the latter could be used to make glass or hydrolyzed to yield lime and silica. The reactions involved in this process are given in

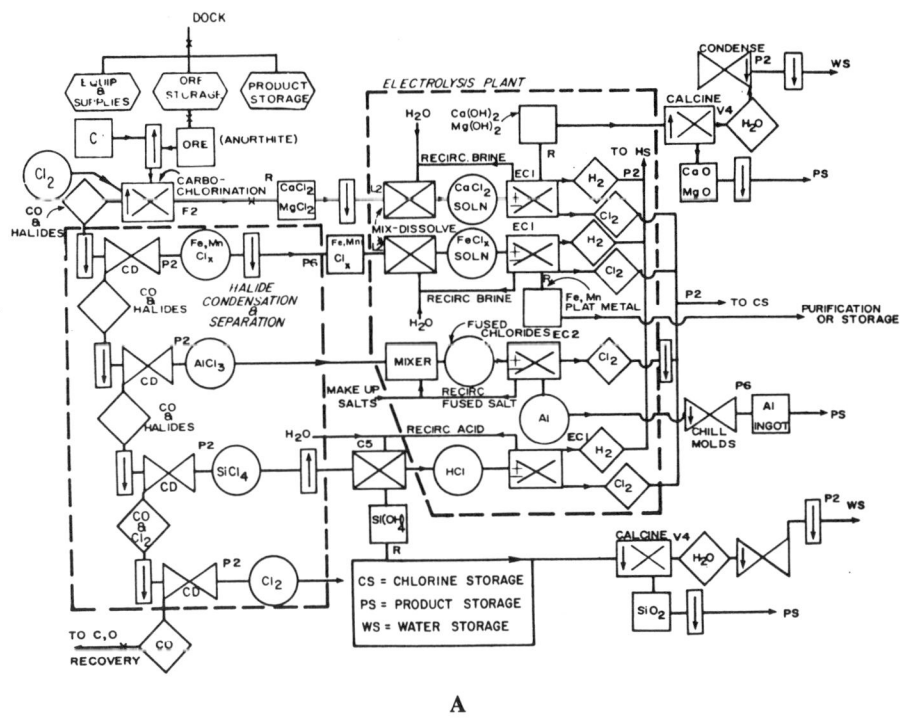

A

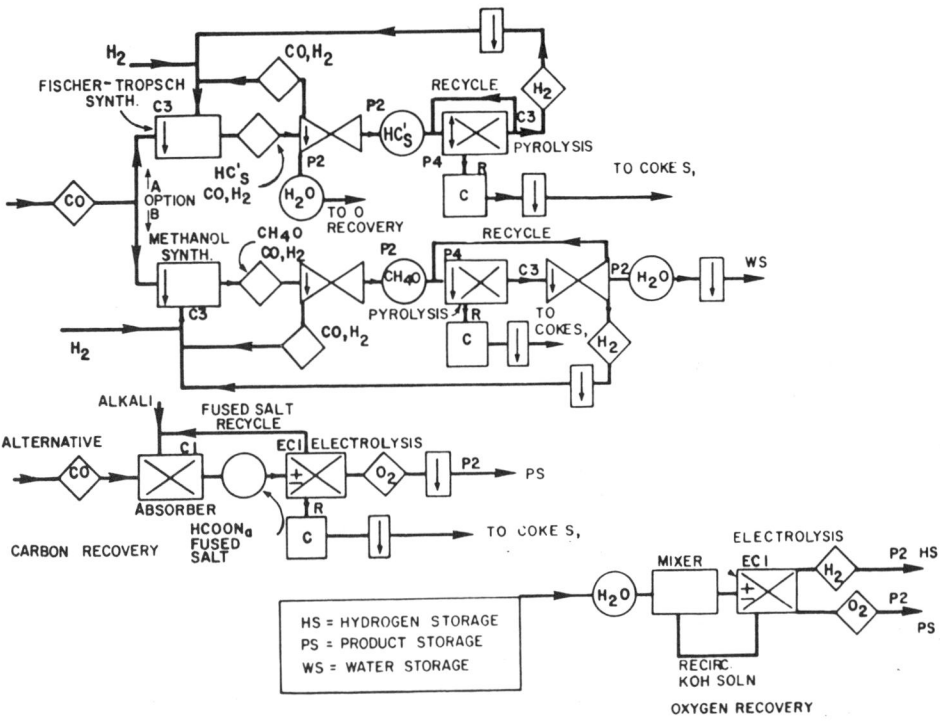

B

FIGURE 5. Carbo-Chlorination process flow chart. (A) Chlorination and separation, (B) reagent regeneration. (From NASA, Lyndon B. Johnson Space Center, Houston.)

Table 9
NaOH BASIC LEACH PROCESS EQUATIONS

(1) $2(CaO \cdot Al_2O_3 \cdot 2\ SiO_2) + 6\ NaOH + 2\ H_2O \xrightarrow[\text{autoclave}]{300°C}$

$Na_2O \cdot Al_2O_3 + Na_2O \cdot 2\ CaO \cdot 2\ SiO_2 \cdot H_2O + Na_2O \cdot Al_2O_3 \cdot 2\ SiO_2 \cdot H_2O$

(2) $Na_2O \cdot Al_2O_3 \cdot 2\ SiO_2 \cdot H_2O + 2\ NaOH + 2\ CaO + 2\ H_2O$

$\xrightarrow[\text{slurry}]{220°C} Na_2O \cdot Al_2O_3 \cdot 3\ H_2O + Na_2O \cdot 2\ CaO \cdot 2\ SiO_2 \cdot H_2O$

(3) $Na_2O \cdot 2\ CaO \cdot 2\ SiO_2 \cdot H_2O \xrightarrow{95°C} 2\ (CaO \cdot SiO_2 \cdot H_2O) + 2\ NaOH$

(4) $CaO \cdot SiO_2 \cdot H_2O + 2\ H_2O \rightleftharpoons Ca(OH)_2 + Si(OH)_4$

(5) $3\ H_2O + Na_2O \cdot Al_2O_3 + CO_2 \xrightarrow{25°C} 2\ Al(OH)_3 + Na_2CO_3$

(6) $2\ Al(OH)_3 \xrightarrow[\text{calciner}]{1100°C} Al_2O_3 + 3\ H_2O$

(7) $Na_2CO_3 + CaO + H_2O \rightarrow 2\ NaOH + CaCO_3 \downarrow$

(8) $CO_3^= + 2\ R^*OH \rightarrow R_2^*CO_3 + 2\ OH^-$

(9a) $CO_3^= \rightarrow CO_2 \uparrow + 1/2\ O_2 \uparrow + 2\ e$

(9b) $Na^+ + H_2O + e \rightarrow Na^+ + OH^- + 1/2\ H_2 \uparrow$

(10) $CaCO_3 \xrightarrow{\Delta} CaO + CO_2$

(11) $R_2^*CO_3 + 2\ HX \rightarrow 2\ R^*X + CO_2 \uparrow + H_2O$

(12) $2\ R^*X + CaO + H_2O \rightleftharpoons CaX_2 + 2\ R^*OH$

Note: R^* = ion-exchange resin.

Table 9. For both basic and acidic leaching, sodium present in lunar soil can probably make up for any sodium lost during recycling. In this process, calcium impurities in the recycled NaOH would not present a problem as base, and not pure NaOH, is needed. A flow diagram for this process is shown in Figure 6.

E. HF Acid Leach Process

This process uses low temperature hydrochemical (hydrometallurgical) steps to separate the silica content of the lunar raw material from the other metallic oxides by conversion to fluorides and fluosilicates followed by vaporization of the silica as SiF_4, and separation of the calcium and the structural metals (Al, Fe, Mg, Ti) by a variety of solution, precipitation, ion-exchange, or electrolytic steps. Generalized process equations are given in Table 10. Iron may easily be recovered from solutions by electrowinning, but the remaining metals except Mg are preferably recovered by sodium reduction of the corresponding fluorides, fluosilicates, or fluoaluminates. Magnesium may be made by silicon reduction of MgO.

Sodium for the metals and silicon reduction can be conveniently obtained by a slight modification of the Castner cell which at one time was the major commercial method for producing sodium. The Castner cell uses the electrolysis of molten NaOH to produce Na, O_2 and H_2. For lunar operations, the hydrogen is an undesirable by-product which can be largely eliminated by using a diaphragm cell and vacuum drying the anolyte to remove the water formed by discharge of OH^- ions.

Metal oxides and silica are obtained, where desired, by hydrolysis of the corresponding fluorides or fluosilicates with steam (or with NH_3 if desired for SiO_2) or by ion-exchange (or permeation) methods. Detailed analyses of the options available for these separations remain to be completed.

A pictorial flow diagram for the HF acid leach process is shown in Figure 7.

Of the processes studied to date, the HF acid-leach one appears to have the best potential for minimal operating mass, ease of element separations to high purity, flexibility, and favorable energy and heat-rejection requirements.

SEPARATION

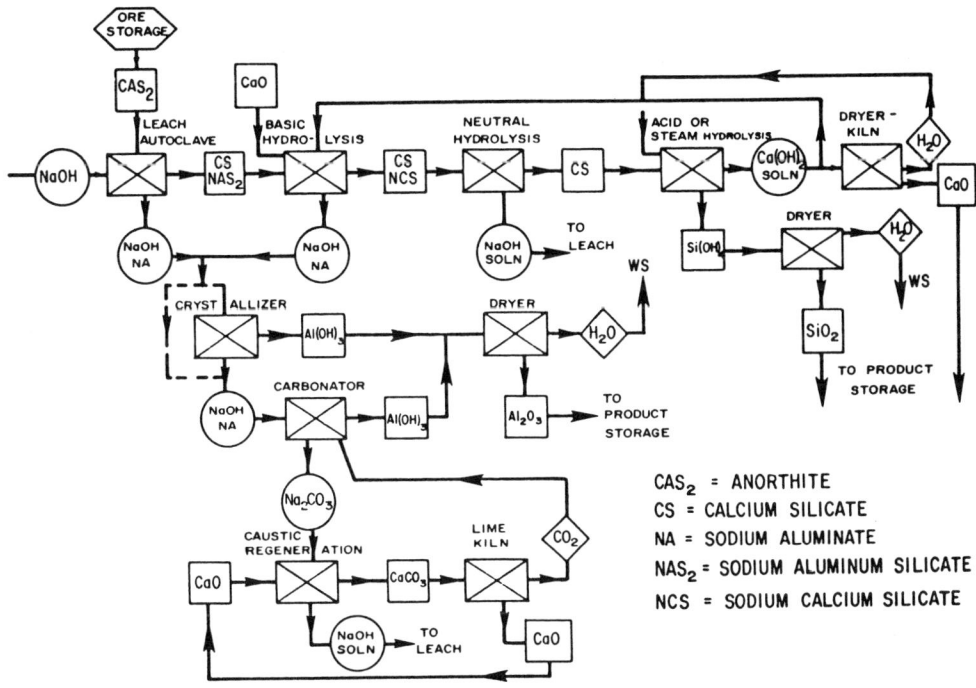

FIGURE 6.   NaOH basic leach process flow chart. (From NASA, Lyndon B. Johnson Space Center, Houston.

Table 10
HF ACID LEACH PROCESS EQUATIONS

(1) $xMO \cdot SiO_2 + (4 + 2x) HF = xMF_2 + SiF_4 \text{ (aq)} + (2 + x) H_2O$

(1a) $xMO \cdot SiO_2 + (5 + 2x) HF = xMF_2 + HSiF_5 \text{ (aq)} + (2 + x) H_2O$

(2) $SiF_4 \text{ (aq)} + NH_2O = SiF_4 \text{ (v)} + nH_2O(v)$

(2a) $HSiF_5 \text{ (aq)} + nH_2O = SiF_4 \text{ (v)} + HF \text{ (aq)} + nH_2O(v)$

(3) $(1-y) [SiF_4 \text{ (v)} + 4H_2O = Si (OH)_4 + 4 HF]$

(3a) $(1-y) [SiF_4 \text{ (v)} + 2H_2O = SiO_2 + 4HF]$

(4) $(1-y'z) [xMF_2 + H_2O = xMO + 2xHF]$

(5) $y [SiF_4 + 4Na = Si + 4NaF]$

(6) $y'[xMF_2 + 2xNa = xM + 2xNaF]$

(7) $z[xMF_2 + xSiF_4 \text{ (aq)} = xMSiF_6 \text{ (aq)}]$

(8) $z[xMSiF_6 \text{ (aq)} + xH_2O + \text{electrical energy} = (x/2)O_2 + xM + xH_2SiF_6]$

(8a) $z[xMSiF_6 \text{ (aq)} + M'SO_3R^* = xM'SiF_6(aq) + xMSO_3R^*]$

(9) $mNaF + mR^*OH = mNaOH + mR^*F$

(9a) $mNaF + (m/2) Ca (OH)_2 = mNaOH +)(m/w) CaF_2$

(10) $mNaOH + \text{electrical energy} = mNa + (m/4)O_2 + (m/2)H_2O$

(11) $(1-y) [Si (OH)_4 = SiO_2 + 2H_2O]$

Note: $R^* = \text{ion-exchange}; m = 4y + 2xy'$

A number of details and options remain to be investigated. Many of these specific process steps can be fully defined by straightforward experiments.

Separation of the fluoro compounds of the metallic elements by solubilities of fluorides or fluosilicates as a funtion of pH and F:Si ratio—with or without additional ion–exchange or electrolytic steps—will require extended literature searches and laboratory investigations. Pyrolytic and hydrolytic behavior of fluorides, fluosilicates, and fluotitanates will also require additional research.

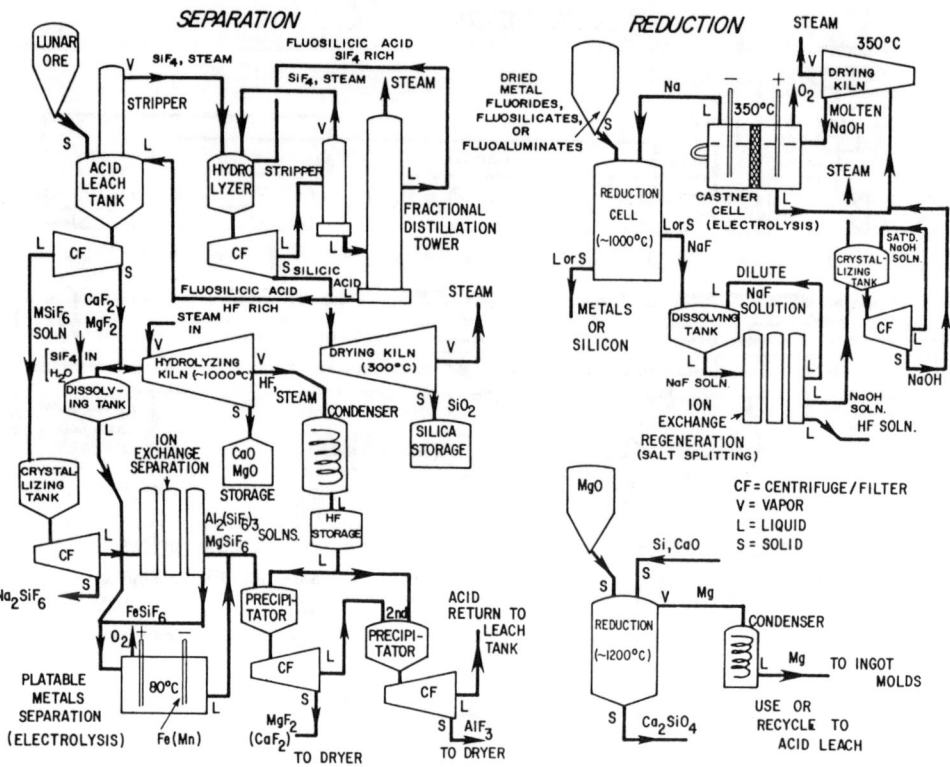

FIGURE 7.   Pictorial flow chart; HF acid leach process. (From NASA, Lyndon B. Johnson Space Center, Houston.)

Despite these informational gaps, nearly all of the proposed operating steps have been studied on a laboratory scale, and about 75% of the steps have been conducted on a pilot or commercial scale under equivalent or comparable conditions.

1. Thermochemistry

The heat transfer requirements of the HF acid leach process may be derived from existing thermochemical data for the compounds present in the process equations. These may be used to prepare the ΔH vs T map shown in Figure 8. In this figure the enthalpy changes involved in water transfer—distillation and condensation—have been omitted. The electrolysis heat load represents only the ohmic heat loss of the process step.

The total input power requirement is projected to total 4100 kWh/t. For the plant sizing analysis, this was increased to 7130 kWh/t to allow for various losses.

2. Plant Sizing

The techniques of plant sizing and mass estimation developed in the preceding discussion were applied to the HF acid leach process. The results are shown in Table 11. By summing the equivalent times for containers, net masses of lunar deficient elements and extra inventory and equipment, one obtains a total equivalent time of 81 hrs. If one uses gross instead of net reagent mass, the total becomes 155 hrs. In either event, the plant should be able to process more lunar ore than its total mass each week, or on an annual basis it could supply more than 50 times the Earth lift mass of the plant.

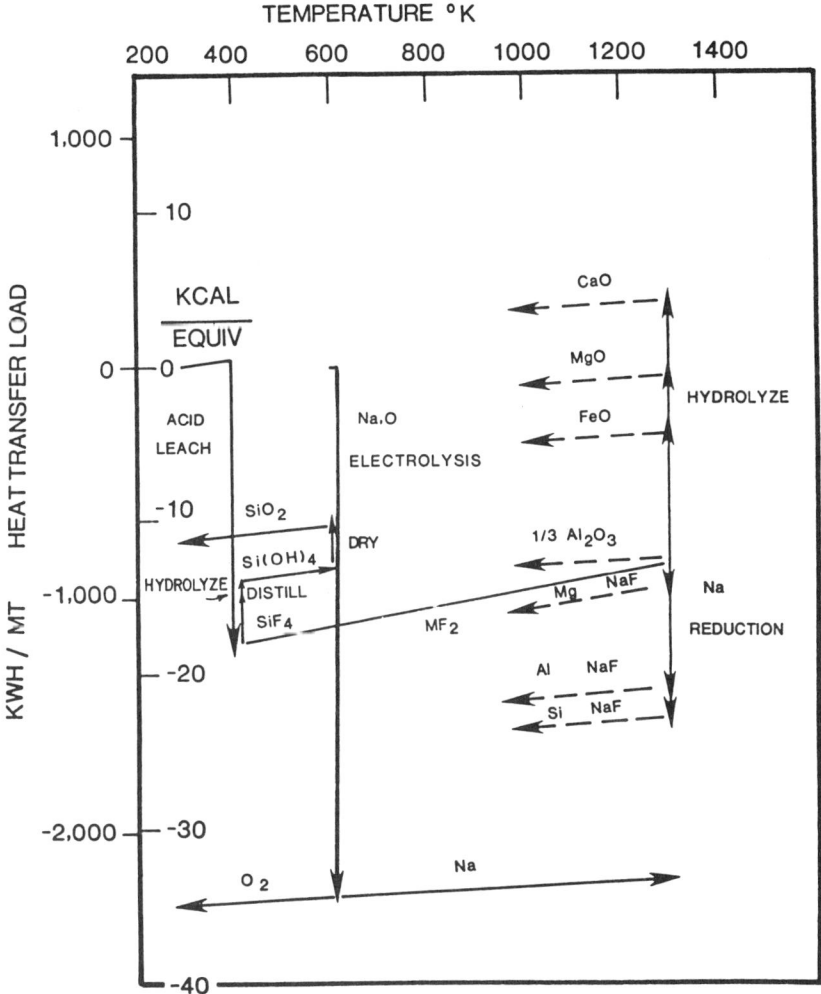

FIGURE 8.   Heat transfer vs. temperature; HF acid leach process. (From NASA, Lyndon B. Johnson Space Center, Houston.)

3. Reagent Replacement Mass

The principal lunar deficient elements used in the HF acid leach process are H, F, Na and optionally N. (Sodium is probably not a serious consideration since it is a minor lunar element and frequently occurs at levels exceeding 2.5% of the total metallic equivalents in feldspar fractions from mare soils.) These recycling elements will occur principally as chemically or physically combined water or hydroxyl ion (H) or fluoride ion (F) and as ammonia or ammonium ion (N).

The hydrogen and nitrogen content of vacuum dried, calcined refractory oxides, or other compounds can be reduced to almost any desired level given sufficient time and temperature at low pressure although mass efficient operation may not be served by drying to less than 0.1% as discussed earlier. If we assume a conservative residual content of 0.5% H_2O and 0.1% NH_3 for the nonmetallic output streams and zero for these impurities in metal outputs, the ratios of reagent replacement mass:input mass for H and N may be expected to fall in the ranges 3.7 to 7.4 $\times$ 10^{-4} amd 4.1 to 8.2 $\times$ 10^{-4} depend on amount of reduced products. The moisture content or dew point of oxygen can be held to insignificant levels.

Table 11
EQUIPMENT AND REAGENT MASSES—PROCESS AND EQUIVALENT TIMES

Step	t Process time (hr)	R_m Contents (equiv. input)	r_m Mass ratios container (equiv. input)	R'_m Net contents (LDE) (equiv. input)	$h_1 = tR_m$ Contents	$h_2 = tr_m$ Container	$h_3 = tR'_m$ Net (LDE) contents
Acid leach	0.5	22.2	0.48	7.95	11.1	0.24	3.97
Sedim centrifuge/							
Distill	0.0167	22.2	28.8	7.95	.371	0.48	0.133
Hydrolyze	0.5	44.9	0.96	14.7	22.5	0.48	7.36
Sedim. centrif.	0.0167	44.9	42.6	14.7	0.75	0.71	0.207
Distill/condense	0.00278	7.06	86.4	2.14	0.0196	0.24	0.006
Distill (½)	0.00278	22.5	86.4	7.36	0.0625	0.24	0.020
Hydrolyze/dry	3.0	3.0	0.55	0.728	9.0	1.66	2.18
Distill	0.00278	7.63	86.4	1.36	0.021	0.24	0.004
Electrolysis	27.5	1.45	0.173	0.036	40.05	4.75	1.0
Metals reduction	3.0	1.76	0.48	0.7	5.28	1.43	2.1
Regeneration(est)	0.5	5.0	0.48	0.7	2.5	0.24	0.35
Misc.	0.5	1.0	0.4	1.0	0.5	0.2	0.5
Subtotal	35.54				92.15	10.91	17.83

Equivalent times* (hr)

Extra mass   (Metric tons @ 4.21 ton/hr input)   Equiv. time (hr)

Reagent inventory	20	4.7
Compressors	10	2.4
Heat exchangers	10	2.4
Pipes, valves	5	1.2
Electrical	6	1.4
Structural & misc.	25	5.94
Radiators (20 MW)	24	5.7
Elec. power (30 MW)	120	28.5
Subtotal	220	52.24

The residual fluorine level will occur principally as residue from steam hydrolysis of refractory fluorides, and cannot be baked out in a practical manner. Analytical studies[23] suggest that the fluorine content of pyrohydrolyzed fluorides can be reduced to 0.25 to 0.5% without undue extension of the process. Vacuum cycling may possibly lead to lower residual levels. We may anticipate F replacement mass:input mass ratios in the range of 1 to 2 × 10^{-3} and possibly lower with improved vacuum desorption.

Total reagent replacement mass:input mass can probably be held below 1 part in 300 with further improvements likely. At this rate, the reagent replacement mass requirement would equal the initial net plant mass in 3 to 6 years. It may be noted that replacement mass need not be supplied in elemental, toxic, or hazardous forms.

4. Plant Scaling

The mass and volume of the processing facility are expected to scale almost linearly with annual throughput. Minimum practical installations may range as low as 1 kg/hr (7 to 8 t/yr). At a very small scale, the mass:throughput and power:throughput ratios may be expected to increase, perhaps by as much as a factor of two.

The concept of the bootstrap growth of processing capacity can increase the annual output mass:net plant mass ratio by an amount limited by the reagent requirements for lunar deficient elements. For the HF acid leach process, the LDE net reagent mass represents 20% of the net plant mass, so the annual output:net plant mass ratio could be nearly quadrupled by expanding equipment capacity using lunar materials. Even greater increases may result from process revisions or modifications of operating cycles which can reduce the equivalent times required for LDE reagents.

ACKNOWLEDGMENT

The Lunar and Planetary Institute is operated by the Universities Space Research Association under Contract No. NSR 09-051-001 with the National Aeronautics and Space Administration. This paper is LPI Contribution No. 379.

REFERENCES

1. **Williams, R.,** Ed., *Handbook of Lunar Materials,* Lunar and Planetary Sciences Division, Johnson Space Center, Houston, 1978, 187.
2a. **Gaffey, M. J., Helin, E. F., and O'Leary, B.,** An assessment of near-earth asteroid resources, *Space Resour. Space Settlements,* NASA SP-428, U.S. Government Printing Office, Washington, D.C., 1979, 191.
2b. **O'Leary, B., Gaffey, M. J., Ross, D. J., and Salkeld, R.,** Retrieval of asteroid materials, Space Resour. Space Settlements, NASA SP-428, U.S. Government Printing Office, Washington, D.C., 173, 1979.
2c. **Bender, D. F., Dunbar, R. S., and Ross, D. J.,** Round-trip missions to low-Delta-V asteroids and implications for material retrieval, Space Resour. Space Settlements, NASA-SP-428, U.S. Government Printing Office, Washington, D.C., 161, 1979.
3. **Gaffey, M. J. and McCord, T. B.,** Asteroid surface materials: A mineralogical characterization from reflectance spectra, *Space Sci. Rev.* 21, 555, 1978.
4. **Brenner, A.,** Alloy electrodeposition, in *Encyclopedia of Electrochemistry,* Hampel, C., Ed., Reinhold, New York, 1964, 21.
5. **Lewis, R.,** Aluminum electrowinning, in *Encyclopedia of Electrochemistry,* Hampel, C., Ed., Reinhold, New York, 1964, 32.

6. **Demmerle, R.,** Castner, Hamilton Young, in *Encyclopedia of Electrochemistry,* Hampel, C., Ed., Reinhold, New York, 1964, 154.
7. **Chilton, C.,** Ed., *Chemical Engineer's Handbook,* 4th ed., McGraw-Hill, New York, 1973, 20.
8. **Chapman, A.,** *Heat Transfer,* MacMillan, New York, 1960.
9. **Dooling, D.,** Space factories in 1997 (Part 1), *Spaceflight 19,* 1977, 342.
10. **Comenetz, G., and Kelly, J.C.R., Jr.,** Containment of hot matter, in *High Temperature Materials and Technology,* Campbell, I. E. and Sherwood, E. M., Eds., John Wiley & Sons, New York, 1967, chap. 19.
11. **Post, R. F.,** Nuclear fusion in *Ann. Rev. of Energy,* Hollander, J. M. and Simmons, M. K., Eds., Ann. Reviews, Inc., Palo Alto, Calif., 1976, 213.
12. **Mager, A. and Brenner, R.,** *Z. Angew. Phys.,* 10, 36, 1958.
13a. **Kesterke, D. G.,** Electrowinning of oxygen from silicate rocks, RI 7587, U. S. Department Interior Bureau of Mines, Washington, D.C., 1971.
13b. **Oppenheim, M.,** *Mineralogical Mag.,* 37, 568, 1970.
14. **Phinney, W. C. Criswell, D. R. Drexler, E., and Garmirian, J.,** Lunar resources and their utilization, in *Space-based Manufacturing from Non-terrestrial Materials,* O'Neill, G., Ed., American Institute of Aeronautics and Astronautics, New York, 1977, chap. 3.
15. **Vachet, P.,** Aluminum and aluminum alloys, in *Encyclopedia of Chemical Technology,* Vol. 1, 2nd ed., Kirk, R. E. and Othmer, D. F., Eds., John Wiley & Sons, New York, 1963, 929.
16. **Rao, D. B., Choudary, U. V., Erstfeld, T. E., Williams, R. J., and Chang, Y. A.,** Extraction processes for the production of aluminum, titanium, iron, magnesium, and oxygen from non-terrestrial sources, *Space Resour. Space Settlements,* NASA-428, U.S. Government Printing Office, Washington, D.C. 1979, 257.
17. **Williams, R., McKay, D., Giles, D., and Bunch, T. E.,** Mining and beneficiation of lunar ores, *Space Resour. Space Settlements,* NASA-428, U.S. Government Printing Office, Washington, D.C. 1979, 275.
18. **J. M. Daubenspeck and C. L. Schmidt,** U.S. Patent 2,852,362, 1958.
19. **Rakhimov, A. R., Ponomarev, V. D., and Ni., L. P. Izv.** Vysshikh Ucheben. Zanvendenii, *Tsvetn. Met.,* 6, 111, 1963.
20. **Shpigun, A. A., Sazhin, U. S., Fedorenko, Y. G., and Shor, O. I.,** *Zh. Prikl. Khim,* 42, 277, 1969.
21a. **Ni, L. P., Medvedkov, B. E., and Ponomarev, V. D.,** *Met i Khim Prom. Kasakhstana Nauch. Techn. Sb.,* No. 3, 34, 1961.
21b. **Ni, L. P., Osipova, E. F., Bunchuk, L. V., and Ponomarev, V. D.,** *Met i Khim Prom. Kazakhstana Nauch. Tekhn. Sb.,* No. 6, 32 1961.
22. **Angstadt, R. L., and Bell, R. N.,** U.S. Patent 3,642,437, 1972.
23. **Silverman, H. P., and Bowen, F. J.,** *Anal. Chem.,* 31, 1960, 1959.

Chapter 6

ASTEROIDAL RESOURCES FOR SPACE MANUFACTURING

Brian O'Leary

TABLE OF CONTENTS

I. INTRODUCTION

For a number of years, the potential of asteroids as resources or for space habitats has been a subject of interest to several visionary scientists and writers.[1] In more recent years, O'Neill proposed that lunar and asteroidal materials may be economically mined and processed in space for the construction of high orbital habitats[2] and satellite solar power stations.[3] The 1976 and 1977 NASA-Ames Summer Studies on Space Settlements[4,5] corroborated early estimates that these concepts could be carried out on a large scale in the 1990s with an initial investment comparable to that of the Apollo program using present technologies. In the scenario considered, an electromagnetic mass driver on the moon would propel lunar material into free space for subsequent transfer to a space manufacturing facility (SMF) for chemical processing and fabrication into large structures. The mass driver would also serve an as orbital transfer vehicle, possibly reducing the cost of transport from low Earth orbit, where the external tanks for the Space Shuttle would be pelletized into reaction mass.[6]

O'Leary[7] explored the possibility of using mass-driver tugs to move Earth-approaching asteroids at opportunities of low-velocity increment to the vicinity of the Earth. Carbon, hydrogen, nitrogen, and free metals, apparently scarce on the moon, may be abundant on some asteroids; possibly, the retrieval of asteroidal materials may be cost-competitive with that of lunar materials in an early program of space manufacturing. A scenario was developed[8] where a 100-MW mass driver, assembled in space with about 50 Space Shuttle flights, would retrieve about 22% of a 200-m-diameter (10-million-t) asteroid through a velocity increment, ΔV, of 3 km/sec in 5 years. Many such objects are believed to be within reach of Earth-based telescopes in ongoing search programs.

The most thorough analysis of the potential of asteroidal resources was carried out at two NASA-sponsored workshops[9,10] held during the summer of 1977. Both studies confirmed the earlier studies that selected Apollo and Amor (Earth-approaching) asteroids could provide cost-competitive resources for space manufacturing, as well as being scientifically interesting targets. Their accessibility (low velocity increment Δv), their zero gravity fields, full-time intense sunshine for energy, and the likely availability of water, carbon, nitrogen, and free metals form a combination of features unique to the Earth-approaching asteroids.

Although funding limitations have necessitated a lower level of effort since the summer of 1977, a number of recent developments, derived from recent observations, theoretical studies and new ideas, tend to make the asteroid option more attractive than was assumed during the summer studies. Details of the 1977 studies and newer concepts are reviewed in this chapter.

In this chapter, scenario development for asteroid retrieval will be emphasized. During the 1977 NASA-Ames summer study on asteroidal resources, two companion papers described in detail: (1) Δv requirements for favorable round-trip missions to currently known candidates and probable future ones,[11] and (2) asteroidal resources and recommendations for expanding the search program, follow-up for orbital determination and chemical classification, and identification of precursor missions.[12]

The first portion of this chapter describes scenarios for asteroid retrieval (1) for a real object (asteroid 1977 HB with gravity assists from Earth, Venus, and the moon) and (2) for a likely hypothetical case, given an increase in the asteroid discovery rate and improved mission-analysis techniques. Several topics are discussed: values for Δv, energy requirements, role of man, design of outbound mass-driver system, assembly and attachment to asteroid, mining operations, timing and logistics of the retrieval operation, the selection of volatiles and free metals, and options for using hydrogen and oxygen processed at the asteroid for fuel. The chapter then places this scenario in a parametric context, which will identify the most significant variables in comparing the economics of transport, into a stable high orbit, of asteroidal, lunar, and terrestrial materials. The next section explores special concepts such as double lunar gravity assists, inter-asteroid collisional velocity change, atmospheric grazing reentry velocity change, and the potential of the moon of Mars, Phobos, as a resource.[13] The chapter concludes with recommendations for a research and development program designed to provide technology readiness for asteroid retrieval by the mid 1980s.

I appreciate the help and advice of many individuals who have contributed significant insight into the development of these concepts: Michael Gaffey, Robert Salkeld, David Ross, David Bender, Clifford Singer, and Brian von Herzen.

II. GROUND RULES AND ASSUMPTIONS

First, I assume that it is desirable to minimize the mass of the equipment required to retrieve the asteroid: rendezvous engine and propellants, mining apparatus, equipment for preparing reaction mass at the asteroid, and equipment for successfully returning the asteroid (or portions thereof). The mass of the retriever must be kept low because of the large Earth-to-orbit launch costs (at about \$700/kg and \$240/kg for an upgraded Space Shuttle) and equipment costs (assumed to be \$400/kg or three times that of a modern aircraft).

The selection of an optimal exhaust velocity, V_e, for the asteroid-retriever mass driver depends on a number of parameters. The earlier asteroid-retrieval study[8] considered V_e ~ 2 km/sec, which was optimized for maximum returned mass per unit power imparted to the reaction mass through a total one-way Δv of 3 km/sec, a value that appears reasonable for the most favorable known cases.[11] This value suggested commonality in design with the lunar mass driver and mass drivers designed for orbital transfer between high orbits (e.g., long-term SMF orbits, lunar orbit) and geosynchronous orbit. A higher-thrust mass driver would act as a booster to spiral the asteroid retriever from low-Earth orbit to high-Earth orbit, as in the lunar scenario.[6]

However, further consideration of the relationship between exhaust velocity and mass-driver mass for constant thrust and an analysis of the requirements for the outbound leg of the retrieval mission led to the conclusion that a single-stage, high-thrust mass driver alone may perform the retrieval. For a nominal Earth escape of Δv of 3 km/sec, in addition to the 6.4 km/sec required to achieve escape from low-Earth orbit by low thrust, V_e ~ 8 km/sec appears to be a reasonable value for the mass-driver retriever. Commonality in design is again possible when the former booster becomes the retriever itself. These assumptions may become invalid if we consider mass-drivers constructed in high orbit, where the total outbound Δv to asteroid rendezvous could be on the order of 1 to 3 km/sec. Nevertheless, the assumptions provide a conservative baseline case.

The use of the same mass-driver design for diverse tasks suggests minimal development costs which would be uniquely attributable to the asteroid retriever. We have assumed that a mass-driver would be used for the first asteroid retrieval because of the common design property and because the mass-driver has some attractive features not offered by most alternative propulsion systems: high throughput-to-mass ratio, relative ease of preparation and use of asteroidal material as reaction mass (the fuel is "free" at the asteroid), high efficiency of operations, and near-term technology required for timely development.

I have further assumed that the size of the asteroid fragment to be retrieved is about 100 m in diameter (1 million t). Such a fragment is at the lower end of the size range of objects accessible to telescopic search programs and it would provide about 0.5 million t of material in high-Earth orbit, which is comparable to a few years throughput of lunar material in the lunar resource retrieval scenario.[3,4] This size would be appropriate for first mission(s) where "one-time" development costs would be in reasonable proportion to total mission costs while their sum would be within the scope of current NASA planning and would require moderate, temporary use of an upgraded (class II) Space Shuttle. A more quantitative investigation of this assumption is presented later. If the best initial candidate is more massive than 10^6 t, it would be necessary to fragment a piece for the first recovery.

I assume that the first asteroid retrieval mission will be manned. In my opinion, the assembly and mining operations are sufficiently complex, with real-time decisions governing equipment several light-minutes from Earth, that an automated mission would be a formidable task. It follows that the total mission time should be kept as low as realistically possible, with crew changes only about once a year. Such crew changes could be conveniently carried out at times of Earth gravity assist (described later). A minimum round-trip time for an asteroid mission would be about 1 year, corresponding to a one-way Hohmann transfer time of about 6 months from Earth to the asteroid on an Earthlike orbit. But most optimal transfer times are considerably longer, particularly when gravity assists are used.[11] For the optimal retrieval of a 10^6 t asteroid, a throughput of about 4 kg/sec of reaction mass would be required for a 3-year return time.

III. VOLATILE AND FREE METAL RETRIEVAL FROM ASTEROIDS

Most ordinary and low-grade carbonaceous chondrite meteorities—and, by inference, asteroids—contain $\gtrsim 10\%$ free metals, water, and carbon. Water and carbon dominate this phase in carbonaceous objects of types I and II; metsls dominate in ordinary chondrites. One class of carbonaceous objects has appreciable percentages of all three, comprising about 20% of the body by weight. At a modest additional cost in the mining operations of the asteroid, the free metals can be sifted out and magnetically separated and H_2O and CO_2 can be extracted by heating the asteroidal fragments in a solar furnace to about 600°C. After the H_2O and CO_2 are extracted, the remaining material would serve as reaction mass. This operation would be continuous at the asteroid; corresponding to the nominal throughput of 4 kg/sec for a 3-year mission to return about 0.5 million tons of material.

The SMF site would receive about 100,000 t of free metals (mostly Fe/Ni), 50,000 t of water, 20,000 t of carbon compounds with the remainder for reaction mass, shielding, and processing into oxygen, ceramics, glasses, and metals.

The 1977 Summer Study[5] projected that 3100 people could be in high orbit by 1991 to process, in space, nonterrestrial materials into satellite power stations. If it is assumed that 3 kg per person per day is required for consumables (half of which is water in "wet" food with some additional nitrogen and oxygen to replace airlock leakage) and about 0.5 ton per person of carbon, hydrogen, and oxygen imparted for biomass to establish farming operations in space, it follows that about 30,000 tons of water, carbon compounds, and nitrogen will be required for the first decade of large-sale production of satellite power stations from nonterrestrial materials. Unless trapped volatiles are found in permanently shadowed areas on the moon these materials must come from Earth or asteroids. Processing a 10^6-ton carbonaceous asteroid for volatiles will provide more than enough material for consumables and for establishing space farms.

A plant will be required to process the appropriate alloys and to fabricate them into useful structures, but the power requirements for this are more than an order of magnitude less than those for extracting aluminum from lunar soil.

The relative energy requirements to produce structural elements from lunar ore (aluminum oxide) and from asteroidal ores (iron-nickel metal grains) can be estimated from basic chemical data. The energy necessary to reduce iron oxide to metallic iron in a blast furnace is about 17×10^6 J/kg (metal) and the theoretical lower limit (heat of formation) is about 7.5×10^6 J/kg (metal). The energy needed to melt metallic iron (heat of fusion, about 10 cal/g; heat capacity, about 0.15 cal/g, 300°—1600 K) is about 9 $\times 10^5$ J/kg (metal). The ratio between the energy required to reduce an oxide and to melt the metal per unit of metal is about 8 (independent of efficiency). The heat of formation ratios to produce metal from an oxide for aluminum versus iron is between 2 (Fe_2O_3) and 3 (FeO). Thus the energy required to produce metallic aluminum from lunar oxide ore is 15 to 25 times higher than that needed to melt a metallic asteroidal ore per unit mass of metal.

The actual relative energy requirements will depend on system designs (e.g, melting by means of solar furnace vs. electrolysis and associated electricity-producing efficiency), the number of melt-freeze episodes necessary to refine the iron-nickel to construction grade alloy, and the ratio of metal mass necessary to provide the same structural strength (about two to three times more favorable for aluminum). The economics should also consider trace element recovery from NiFe (Ni, Co, Ge, Pt, etc.) during the refinement process of iron.

If the asteroidal NiFe metal is used, the mass and complexity of the SMF chemical processing plant may be reduced considerably, further reducing the cost of an early program of nonterrestrial material processing. It also appears that virtually all the materials required for SMF operations—consumables, plastics, graphite for radiators, solar reflectors, and possibly germanium for solar cells—would be available. If a lunar program is more feasible at first, the economics of volatile and metal extraction from asteroids may be immediately competitive. If serious problems were to develop with the structural dynamics of mass drivers, water could possibly be used at the asteroid as fuel (liquid oxygen and hydrogen produced from electrolysis) or as a working fluid to return appreciable quantities of asteroidal materials. Details of these concepts are explored later.

IV. SCENARIO FOR ASTEROID RETRIEVAL

A representative scenario was developed for capturing a 10^6-ton asteroid[10] (see Figure 1). A single mass driver is used to accelerate the crew and the equipment for mining and retrieval to rendezvous with the asteroid. That mass driver is then used to return a large fragment of the asteroid to Earth orbit. The scenario chosen uses a low-thrust Δv of 6.4 km/sec to achieve Earth-escape speed, a lunar gravity assist maneuver, and a further Δv of 3 km/sec to intersect and rendezvous with the asteroid. Also, in the outbound trip, one or more gravity-assisted maneuvers are used with Earth and perhaps Venus.[11]

Return Δv is taken to be 5 km/sec including insertion into a 2:1 resonant SMF orbit[11] (see Figure 2). A real-case asteroid mission to capture a portion of the asteroid 1977 HB was used as a basic mission.[3] The results of this investigation are shown in Table 1. In this case, which was not optimized but was arrived at by trial and error, the outbound Δv was intolerably large, even with gravity assist. But the return Δv was between 2 and 3 km/sec (the lower value obtainable with an extra Earth gravity assist). The 1977 Ames Summer Study group concluded that, upon further analysis, opportunities will be found for missions to currently known asteroids in which either the inbound or outbound Δv is on the order of 3 km/sec, with the other leg larger. The strategy will be to find opportunities where some combination of both Δv values is minimized. The 5-km/sec return Δv was chosen since lowering the Δv of the initial leg will most likely increase the final-leg

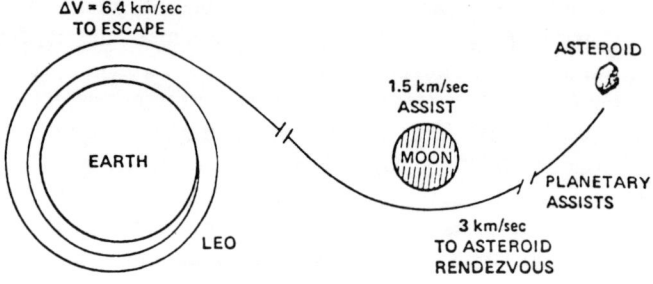

FIGURE 1. Outbound journey.

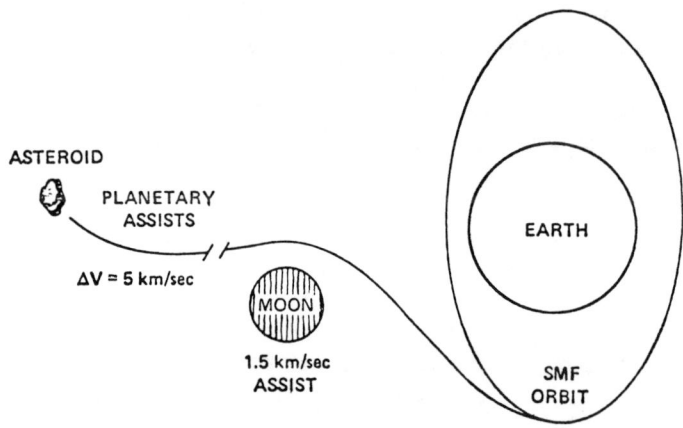

FIGURE 2. Inbound journey.

velocities. As shown in a later section (Parametric Consideration of Asteroid Capture), these values are not critical to mission economics. Since the primary effort of the mission is to expend moving mass out of the gravity well of Earth, the cost per kilogram of returned asteroidal material will increase only slightly if the delta V values are increased by 2 or 3 km/sec. Nevertheless, the best-known opportunities should be identified.

Two considerations strengthen the belief that considerably better missions than that shown in Table 1 are possible. Missions are found by trial and error and, therefore, finding a good mission depends directly on how many cases are considered. Time severely limited the number of cases we analyzed; with more time, the study of known objects can be extended considerably. An increased discovery rate for this type of object would increase the number of good cases and should therefore be pursued. Also, concepts such as velocity change by interasteroid collisions, double lunar gravity assist, and atmospheric graze will further reduce the requirements.

Given these caveats and uncertainties, the scenario is now described. The retrieval of asteroidal material includes the following steps: achieve a low Earth orbit, transfer to high Earth orbit, lunar gravity assist and departure, intermediate gravity-assisted maneuvers, rendezvous operations, including despinning, mining, and processing operations; and mass-driver coupling and deployment; and return. (Each step is described in detail in the following sections.)

A. Low-Earth Orbit

Components of the asteroid-retriever mass driver, fuel for the outbound leg, and mining and processing equipment would be Shuttle-launched from Earth and the

Table 1
RETRIEVAL MISSION TO 1977 HB (TOTAL MISSION TIME, 1974 DAYS)

Location	Operation	Date	Delta V required km/s
Earth	Depart high orbit	April 28, 1984	6.32
1977 HB	Arrive	November 28, 1984	2.53
1977 HB	Depart	July 1, 1985	2.04
Earth	Flyby	April 15, 1987	
Venus	Flyby	March 10, 1988	
Earth	Flyby	August 29, 1988	1.04
Earth	High orbit arrival	September 23, 1989	1.5 (LGA)

component systems should be assembled and tested in low-Earth orbit. If it is assumed that Shuttle payloads are arranged to be mass-limited rather than volume-limited, and if the Shuttle hydrogen tanks (30 t) are used as mass-driver reaction mass, about 55 t may be lifted to 250-km-altitude orbit per launch. When the assembly and testing phases are completed, the crew is Shuttle-lifted to the waiting vehicle. This 21-man team would live in one or more Shuttle hydrogen tanks similar in design to that envisioned for an early space habitat. Escape from near Earth space is achieved in two stages: first, with a Δv of about 6.4 km/sec, the asteroid-retrieval package is raised to high-Earth orbit by a slow spiral orbit. This stage requires 2 weeks to complete, using a mass driver (or its equivalent) with an exhaust velocity of 8 km/sec. The mass driver is optimized to place the asteroid retrieval package in suitable orbit for the minimum number of Shuttle flights, and ideally would be identical to that designed for low-to-high Earth orbit transport.[6] Data on mass drivers generated during the 1977 Ames study indicate that, from conservative assumptions made about the component masses and length, a total mass of 2500 tons is required for an acceleration of 800 gravities. The propellant throughput is 4 kg/sec. The crew and their life-support system have a mass of 291 tons; the mining facility has a mass of 560 tons. A total outbound Δv of 9.4 km/sec requires the launch of 10,000 tons to low Earth orbit. With an assumed return Δv of 5 km/s and an asteroid fragment of 10^6-ton mass, this enables 532,000 tons to be brought to Earth orbit. The second departure phase involves a lunar gravity assist which gives a 1.5-km/sec escape velocity.

B. Gravity-Assisted Maneuvers

Besides the lunar-gravity-assisted maneuver, other possible maneuvers permit a considerable saving in Δv requirements and hence in mass lifted to low-Earth orbit. These are described in reference 11 and are summarized as follows. Of the planets, only Earth and Venus are suitable for missions to Earth-approaching asteroids because the other planets are either too small or too far away. The low-thrust Earth Gravity Assist (EGA) technique uses the Earth alone and is therefore always available. After launch from Earth, the mass driver imparts a small Δv to the system and causes it to reintersect the Earth with a higher relative velocity than that with which it departed. These times of close encounter with the Earth offer excellent opportunities for crew rotation, about every 400 days, except during the trips to and from the asteroid. This relative velocity is changed in direction by the close Earth encounter, and the final sun-centered velocity may be altered greatly.

The second type of gravity-assisted maneuver, called Venus Earth Gravity Assist (VEGA), uses both Earth and Venus. The mission leaves Earth to intercept Venus. A close Venus encounter rotates the relative velocity vector and directs the retrieval package back to Earth. A second flyby of the Earth bends the relative velocity vector and the final sun-centered velocity may be greatly increased. The mass driver may improve conditions further by thrusting during the Earth-Venus or the Venus-Earth

RENDEZVOUS OPERATIONS AT THE ASTEROID

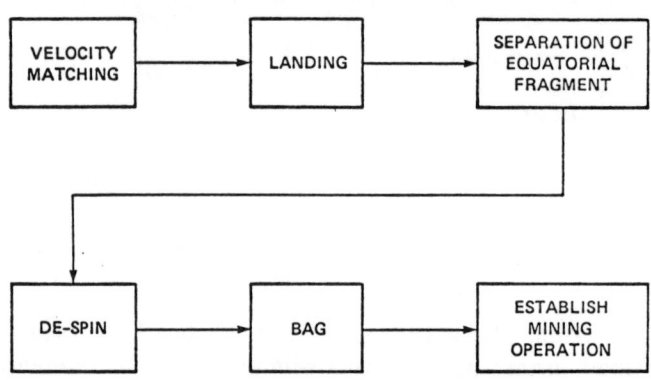

FIGURE 3.   Rendezvous operations at the asteroid.

passage. In all these ways, a considerable saving in Δv and thus in launch mass may accrue. The mission leg from the last planetary encounter will probably require additional thrust to intercept Earth and will certainly require thrust to match orbital velocities. The return trip may use similar gravity-assisted maneuvers to further reduce thrust requirements.

C. Asteroid Rendezvous Operation

Rendezvous with the asteroid includes the following events (Figure 3): orbital matchings, fragmenting, despinning and bagging the asteroid, establishing the mining and processing operation, and packaging material for reaction mass and for return. Orbital matching is achieved in about 1 day, with a total velocity change of approximately 1 to 2 km/sec.

After rendezvous, a small jeep will land on the pole and its crew will find a suitable fragment of the asteroid for return to Earth. This fragment, which will have a mass of approximately 10^6t, will be near the equator and will be separated from the asteroid by a small chemical explosion. The jeep will then land on the detached mass and proceed to despin it. Asteroids rotate at different rates, but 4 revolutions per day may be considered representative of asteroids in the 100-m-diameter size range. A jeep will land at the pole of the fragment and establish a line of small Y-shaped pilons around its equator (Figure 4). These pilons will be either driven into the asteroid (if it is firm enough) or held in place by some kind of anchor. The jeep crew will then wind the asteroid in the direction of its rotation with 5 km of very lightweight cable, using the pilons to hold the cable. One end of this cable will be firmly anchored to the asteroid and the other end to the tug. Launching the jeep with a very minimal velocity to just escape the weak field of the asteroid would result in the cable being unwound as the asteroid rotates. A very tiny thrust from the jeep would be required to keep a light tension on the cable (a few Newtons), which would have a minimal effect on the despin operation. After 2 days, the cable will be unwound and will start to rewind in the other direction. The jeep will thrust against this with 34 Newtons for 4 days (Figure 4). For an exhaust velocity of 4 km/sec, 29 tons of propellant will be used to despin the asteroid completely. Cable and pilons will be stored for return.

After despinning, the asteroid fragment will be bagged along with the entire processing plant. The bag is a slightly pressurized covering constructed of 1-mm

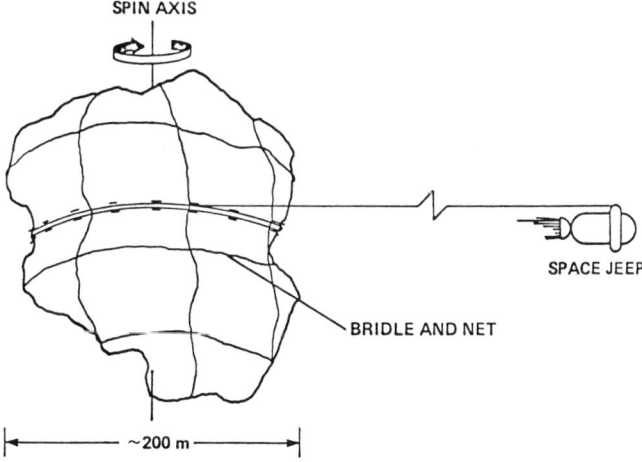

FIGURE 4. Despinning an asteroid.

MINING OPERATIONS

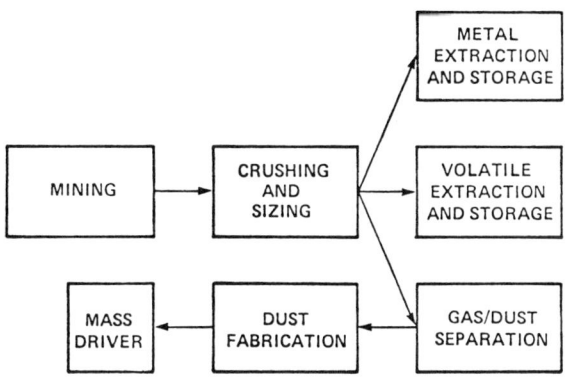

FIGURE 5. Mining operations.

composite material resembling fiberglas plastic-coated to make it gas-tight. This bag, with a volume of $10^6 m^3$, contains debris and gas produced by mining and processing operations and keeps the surrounding region clean. The asteroid fragment itself is surrounded by a net of 1 km² area composed of 0.5-cm aluminum cable spaced 1 m apart. This net will be preassembled and can be rolled out quickly by the crew. The net distributes the thrust loads over the asteroid surface and must be able to withstand the 32 tons of thrust developed by the mass driver. The total mass of the bag and net is 160 tons. The mass driver is now brought alongside the asteroid and coupled to the net. Considerable adjusting will be required to ensure that the mass driver thrusts through the center of mass of the system. Care must be taken in processing the asteroid during return so as not to misalign the thrust by shifting the center of mass.

D. Mining and Processing Operations

After the asteroid has been cradled, tied down, and bagged, the mining and processing operations must be established (Figure 5). The design requirements for the mining and processing equipment will depend on the character of the asteroidal material being used.[12] Three possible types of material containing abundant volatile and/or free metal phases are listed in Table 2.

In this scenario, a metal-rich carbonaceous material is considered to be the most desirable material to recover. We also assume that the metals (NiFe) and volatiles are sufficiently valuable to justify the extra steps necessary to extract or concentrate these materials. There are six major processing steps to follow between the raw asteroidal material and its use as reaction mass in a mass driver: mining, crushing, and sizing; metal extraction and storage; volatile extraction and storage; gas-dust separation; and dust fabrication for mass-driver use. Figure 6 is a schematic diagram of the asteroid processor.

The physical situation at the asteroid requires special types of processing and handling. For an object in the mass range of interest, the surface gravity is about a few microgravities, which corresponds to a surface acceleration of some thousandths of a cm/sec^2 or a few tens of μm/sec.^{2} Initial acceleration proposed for the mass-driver system is about 50 μm/sec^2 and increases as the body mass decreases. A body of this size in the solar system is almost certainly a collisional fragment (or a portion of one) and will be strength-limited by internal fractures. A relatively simple netting arrangement should suffice to hold it together against the relatively small accelerations contemplated. (This netting was described previously.)

1. Mining

The major problem in any mining scenario is to hold the cutting equipment against a surface with sufficient force to cut into the fragment. In tunneling operations, pressure can be exerted on the walls to provide stabilization, but for fractured, relatively low-strength rock, tunnels must be spaced to provide relatively thick walls (e.g., about 1 tunnel diameter), which limits the usefulness of this technique, particularly if much of the material is to be used for reaction mass.

An alternate surface technique involves a covered flail excavator that cuts a 5-m-wide, 1-m-deep trench. This excavator is held against the surface by anchor rods driven into the ground around it; the excavator pulls itself forward with these anchors. The forward progress needed to supply 4 kg/s is about 0.03 cm/s or about 1.3 m/hr. This device delivers appropriately sized fragments (e.g., $\leq$ 20 cm) through a pipe or conveyor system to the main processing facility.

2. Crushing and Sizing

Because the materials being considered are relatively soft, crushing them should be relatively easy. The crushing device used has a rocking jaw arrangement for coarse crushing ($\leq$ 1 cm) and a series of rollers for fine crushing (Figure 7). The crushing elements are arranged radially in a rotating cylindrical housing to provide a radial acceleration from the hub (input) to the final powder ($\sim$ 0.1 g at 10 RPM of a 3-m-radius device). Mean particle sizes of about 0.2 mm or slightly less are desirable for metal extraction.

3. Metal Extraction and Storage

After crushing, it is probably desirable to use gasdynamic transport to move the materials. Gas pressure can be quite low $\sim$ 1 millibar) and still be effective for transport in this low-gravity environment. A low gas pressure simplifies system construction since leak rates and material strength requirements are minimized. The carrier gas could be either CO_2 or H_2O, both of which are available. Leakage into a moderately contained environment (e.g., the surrounding processing facility) can be minimized by cold-trap pumping in a passive mode.

Table 2
MINERALOGICAL, CHEMICAL, AND PHYSICAL PROPERTIES OF THREE POSSIBLE ASTEROIDAL MATERIALS

Type	Metal-rich carbonaceous (~C2)[a]	Matrix-rich carbonaceous (~C1-C2)[b]	Type 3-4, L-H chondrite
Fe[c] (metal)	10.7	~0.1	6—19
Ni (metal)	1.4	—	1—2
Co (metal)	0.11	—	~0.1
C	1.4	1.9—3.0	~0.3
H_2O	5.7	~12.0	~0.5
S	1.3	~2.0	~1.5
FeO	15.4	22.0	~10.0
SiO_2	33.8	28.0	38.0
MgO	23.8	20.0	24.0
Al_2O_3	2.4	2.1	2.1
Na_2O	0.55	~0.3	0.9
K_2O	0.04	0.04	0.1
P_2O_5	0.28	0.23	0.28

Minerals

	Clay mineral matrix Mg olivine with Fe_2O inclusions	Clay mineral matrix olivine	Olivine pyroxene metal
ρ(gm/cm³)	3.3	2.0—2.8	3.5—3.8
Metal grain size	~0.2 mm	—	~0.2 mm
Strength	Moderately friable	Weak-moderately friable	Moderately friable

[a] Data from metal-rich C2 meteorite Renazzo.
[b] Data from C2 meteorite Murchison and average C2-C1 types.
[c] Chemical analysis in weight percent.

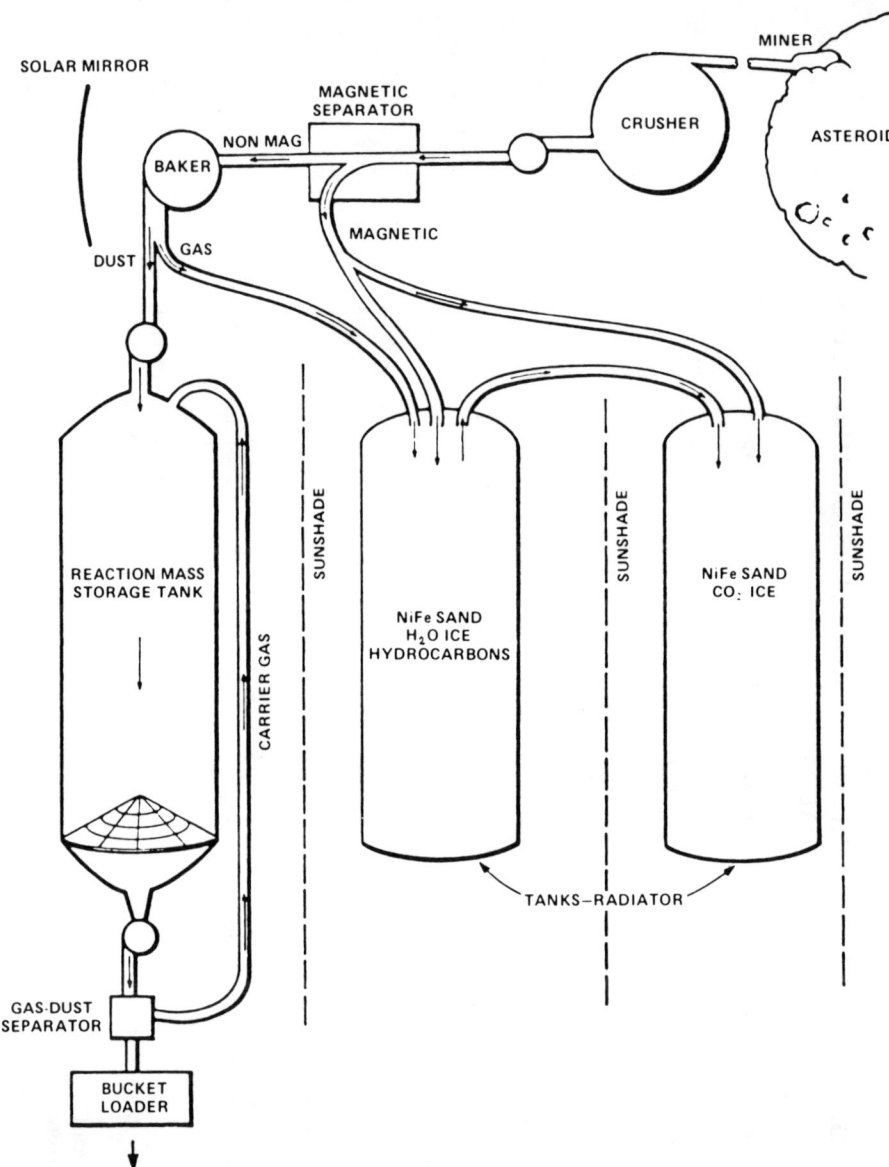

FIGURE 6. Schematic diagram of asteroid processor.

To extract most of the metallic phase, the gas stream and its entrained dust are passed through a magnetic field (Figure 8). An enriched fraction ($\geq 80\%$ NiFe) will contain most of the coarse-grained metal particles ($\sim 70\%$ by weight). This enriched fraction (metal sand) will be diverted to a storage facility (large, relatively weak tank or bag). It is doubtful that any subsequent processing of this metal-rich phase would be economically viable enroute.

4. *Volatile Extraction and Storage*

To recover the volatile phases (H_2O, CO_2, hydrocarbons), the material must be heated (Figure 9). The temperatures required depend on the desired recovery rate. Water chemically bound into the clay minerals of C1-C2 matrix material begins to come

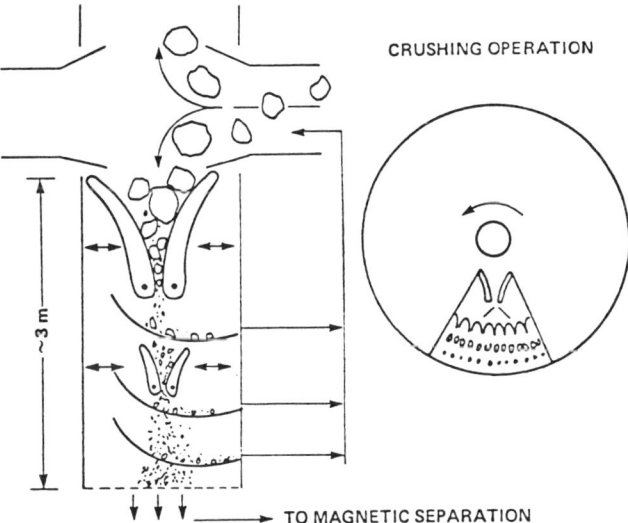

FIGURE 7. Crusher.

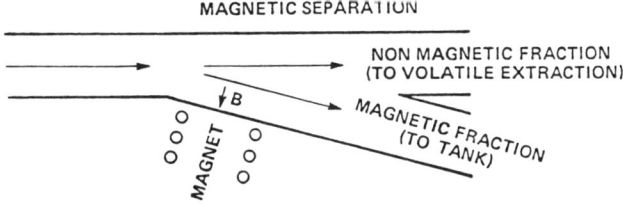

FIGURE 8. Magnetic separation.

off near 100°C and continues to come off until about 400°C. Carbon dioxide is produced at low temperatures ($\sim$ 100 to 200°C) by the breakdown of carbonate minerals (e.g., calcite, etc.) at higher temperatures (400 to 700°C) by the disassociation of hydrocarbon compounds and by reaction of elemental carbon with oxide phases (e.g., $C + 2Fe_2O_3 \rightarrow CO_2 + 4FeO$). Hydrocarbons volatilize or break down in the range 100 to 700°C, releasing a variety of compounds, including methane and petroleum-like vapors.

The energy requirements depend on the recovery rate, which is dominated by the heat of vaporization of water. Considering the various heat capacities of the phases, about 200 cal/g are required to raise raw C_2 carbonaceous matrix material to about 700°C and to vaporize the volatile components. A 4-kg/sec throughput would require a solar collector about 60 m on a side focused on a collector area about 6 m across. Dust entrained in a carrier gas ($\sim$ 2 to 10 kg/m³) is introduced into a heat-exchange system at the mirror focus for a period of 30 to 60 sec. The dust is separated from the gas by a cyclonic separation (dust settles rapidly to the outer wall of a curved conduit). Counterflow of gas and dust is used to heat incoming stream and cool outgoing steam. The gas steam is directed to a shaded heat exchanger and into a storage tank that also serves as a heat exchanger. To condense all products to $\sim$200 K, a radiation area must be about 100 m on a side. The heat released by condensing water may limit the total volatile recovery program. A double condensor system may solve some of this difficulty, the first tank operates near 250 K and precipitates much of the H_2O (depending on vapor pressure) and hydrocarbons. A second tank operating near 150 K

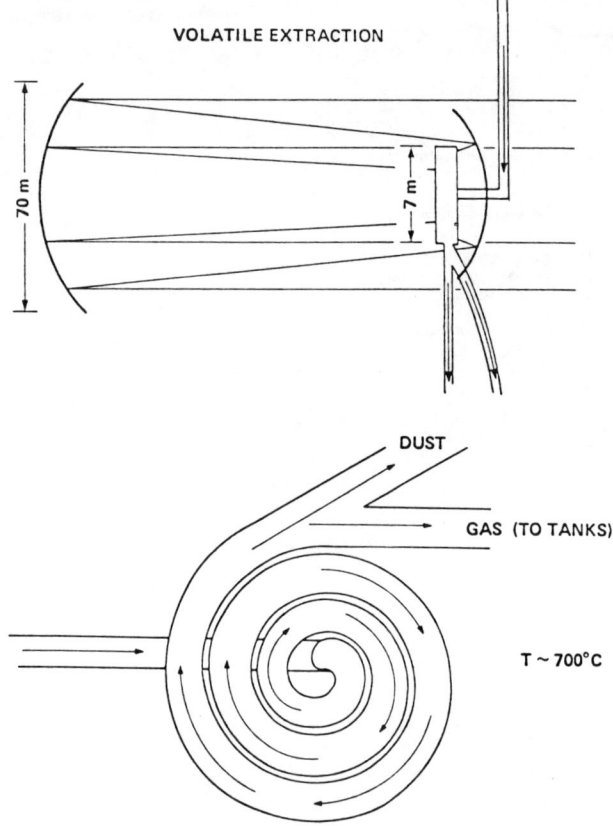

FIGURE 9. High-temperature volatile extraction.

and slightly elevated pressure condenses the remaining H_2O and CO_2 vapors (see Figure 10). The metallic sand is stored in both tanks to increase the thermal conductivity of the ices.

The recovery of volatiles is likely to be limited by the ability to radiate away the heat of condensation of the volatile phases for reasonable tank/radiation mass. In the shade, radiating to a dark sky, an optimum value can be evaluated.

5. Dust Fabrication for the Mass-Driver

The dust fraction of the cooled product is delivered to a storage tank, then to either the mass-driver loading point or to a central fabricating facility. The dust is separated from the entraining gas and is stamped or poured into appropriate molds. The character of these mold requirements is presently unspecified. However, laboratory experience has shown that relatively low pressures are required to form a reasonable solid pellet.

E. Mass Requirements for Mining, Processing, and Storage

The cost and feasibility of mining and processing an asteroid for reaction mass, volatiles, and metals will depend greatly on the mass and associated costs of transporting equipment to the asteroid to accomplish these tasks. The necessary equipment includes that for mining and processing (crushers, metal and volatile extractors), storage containers, radiators, and support equipment (piping, blowers, "nets," and "bags"). An initial estimate of the mass of the necessary equipment can be

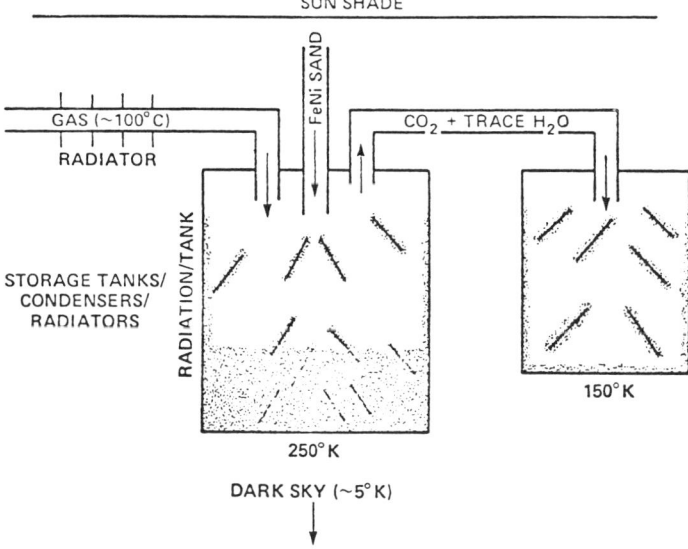

FIGURE 10.   Radiatively cooled volatile storage tanks.

made by fixing a preliminary design for each component and by then making a reasonable estimate as to the mass of that component. Certain elements of the proposed system are sized to a particular mass-flow rate. Each component is discussed briefly below and is associated mass is estimated.

Reaction-mass storage and "pumping" facilities include the system needed to store and feed reaction mass for the outbound leg (analogous to a fuel tank). For a throughput of 4 kg/sec, this tank must have a storage capacity of about 200,000 m³ and is envisioned as a gas-tight (~ 1 mbar pressure) bag, 20 m in diameter and 65 m long, opening slightly "downward" with a gas-agitated grating at the bottom. The bag material is assumed to be a composite (canvas and plastic or appropriate substitute of mass 3.75 kg/m², e.g., 5-mm-thick composite or 2.5-mm-thick aluminum) with a mass fo 25 tons. Gasdynamic transport using blowers (five blowers at 2 tons each with motor), composite pipes (about 1 km, 7 kg/m², and with a 2-m diameter weighing about 22 tons), and a gas-dust separator and bucket loader (about 4 tons and 10 tons, respectively). Such a structure should have a mass of about 71 tons, and if we apply a safety factor, the estimate is 120 tons.

"Big bag" is slightly pressured gas-tight covering for an entire work area (1 mm of composite), which encloses a volume of 10⁶ m³ with a mass of about 60 tons. The asteroid is enclosed and suspended in a net (1 km² of 0.5-cm aluminum cable at 1-m spacing weighing about 100 tons).

The mass of the processing equipment corresponds with the throughput of material. The 50-kg/sec mass flow, considered the maximum case, is discussed here. (For other mass-flow rates, estimates are provided below.) The equipment includes: miners (three at 50 tons), crushers (two at 75 tons), a magnetic separator (10 tons), blowers (five at 2 tons), volatile extraction baker and mirror (200 tons), volatile storage and radiators (two at 350 tons). The total mass requirements depend on throughput as given below: 0 kg/sec—300 tons, 4 kg/sec—560 tons, 10 kg/sec—650 tons, and 50 kg/sec—1500 tons.

F. Alternatives to Mass Driver

Of any presently known system, the mass-driver concept allows an asteroid mission to return to Earth orbit with the largest amount of material. But other devices are available that could also return large quantities of material. For any asteroid recovery mission, it may be preferable to process the material where it is obtained and to bring only the most valuable portion back to Earth. The disadvantages of a longer waiting time at the asteroid may be more than compensated for if processing the asteroid *in situ* allows either a faster manned return or the use of a propulsion system that can be fully automated. The 1977 Summer Study group indicated that establishing machinery to process volatiles and free metals at the asteroid does not impose severe economic penalties.

Two schemes suggest themselves for the return flight: a rocket that uses either water or volatized rock as a working fluid and the solar sail. Solar power could convert water from a carbonaceous asteroid into liquid hydrogen and liquid oxygen by electrolysis. LH_2 and LOX would then be used as fuel and oxidizer in a low-thrust rocket engine that has an extended operating time. Although this is a roundabout way to extract propulsion energy from sunlight, the very high efficiency of the electrolysis makes it feasible. The conversion process would be continuous throughout the return flight, a factor that might enable the return mission to be entirely automated. To avoid the use of large cryogenic storage tanks, the rocket motor would, ideally, also operate continuously on the return flight. This introduces the complication of a long-thrusting rocket motor that uses LH_2 and LOX, but the mass of the rocket motor is negligible compared to either conversion system or payload so that several could be used sequentially.

An in-house General Dynamics study[27] provided the basic figures for the electrolysis and refrigeration plant. It was assumed that, excluding this plant, the mass of the supporting structures and the rocket motor and its subsidiary systems would not together exceed 300 tons. This latter figure appears reasonable since, if the payload is stored in the shadow of the solar collector, it will be solid condensate and should not require an elaborate bracing system to transfer the modest thrust to it. The exhaust velocity, taking into account various losses, is assumed to be 4400 m/sec and all missions have a 1-year return time. The exhaust velocity is compatible with low Δv mission requirements. This type of rocket is well configured for a water-retrieval mission. Changing the mission time directly alters the thrust and collector area requirements, but has only a secondary effect on the mass of the system itself. Processing rates are assumed constant for both outbound and return legs and therefore the thrust, but the rocket operates only during a portion of the outbound leg. Since a conventional rocket may be started or stopped easily, this requirement for noncontinuous thrust on the outbound leg presents no serious complication.

Collector area is based on an assumed 10% solar collector efficiency and a 95% electrolysis efficiency. Both are reasonable, and the possibility of using germanium solar cells with a sunlight concentrator instead of silicon may make the first figure conservative. Collector areas and power requirements are moderate and thrust requirements may be fulfilled with very small engines. For equivalent Δv values up to 8 km/sec, the mission returns more mass from the asteroid than it used to reach the asteroid. This is necessary if the returned material is primarily water but is desirable in any case. All scenarios assume a return payload of 10^4 tons. For low equivalent Δv values, the mass of water in high Earth orbit may be multiplied many times by this device. The same scenario as presented alone is used here, an outbound Δv of 3 km/sec from high Earth orbit and a return of Δv of 5 km/sec from the asteroid. Power supply and electrolysis plant scales inversely with mission return time. For a 2 year return flight, the solar rocket fully fueled in high Earth orbit has a mass of 2084 tons. The rocket has a mass of 1054 tons, 463 tons of which are the power supply and electrolysis plant and 291 tons are the crew and their life-support system. The remaining mass is a processing plant for extracting water and other derived mass from the asteroid. For continous operation, the power requirement is a modest 19 MW. For the return trip in this nonmass-driver scenario, an initial mass of 33,000 tons results in 10,000 tons brought to high-Earth orbit, a tenfold increase in the mass of fuel used for the outbound journey.

A second alternative to the mass driver, one that further reduces the mass brought to Earth by a factor of 10 but which could more easily be automated, is the solar sail. This device consists of a very large mirror of thin metal foil or a microlayer of metal or plastic; it relies on solar radiation pressure—caused by momentum transfer at reflection—for propulsion. Although the maximum force available at Earth orbit is only 9.3 N/km², the need for propellant has been eliminated.

The recent development of new sail material with lightness ratios (ratio of solar radiation pressure to solar gravitational force) of 5 or more has returned the solar sail to consideration for low to moderate Δv low-thrust missions. The lightness ratio is constant regardless of distance from the sun.

It is assumed that the structural mass excluding the sail is 10% of the returned payload. If a 1000 t (10^6 kg) payload must be returned and if it is assumed that 38% of perpendicular thrust can be used, lightness ratios of 5 and 1 both yield satisfactory results. For most missions, the sail area depends very little on sail lightness ratio and therefore the sail area can be varied inversely with mission time.

For specific scenario used previously, a 2-year return time requires a 22.7-km² sail area, a sail mass of 7.2 tons for a sail lightness ratio of 5, and a 23.4-km₃ sail area and a sail mass of 37 tons for a sail lightness ratio of 1. In both cases, 1000 tons of material may be brought to Earth orbit. It these devices are automated, a fleet of these may be launched from a suitable asteroid and later retrieved near Earth. While the mass driver promises to be very useful in returning large quantities of asteroid material to Earth orbit, it is not crucial to such a mission. Either alternative presented here appears feasible and there are surely other possibilities. The solar rocket hybrid is within reach of current technology, while the solar sail is not far beyond it.

V. PARAMETRIC CONSIDERATIONS OF ASTEROID CAPTURE

As part of the 1977 Ames Summer Study, R. Salkeld carried out a parametric comparison between the retrieval of terrestrial, lunar, and asteroidal resources.[10]

The estimated masses of an asteroid mining vehicle sized to capture a 10^6 ton asteroid are shown in Figure 11 as a function of mass-driver exhaust velocity. As exhaust velocity increases, the mass-driver mass increases sharply because of increasing power requirements, while the mass of the mining system including personnel accommodations decreases more slowly because the reaction mass throughput decreases.

It is postulated that the asteroid can be captured by a single-stage mining vehicle starting from LEO, which attaches to the asteroid and returns with it to far Earth orbit. The required start mass in LEO is shown in Figure 12 as a function of exhaust velocity for a typical mission requiring ideal velocity gain on the outbound trip of 9.4 km/sec (6.4 km/sec for escape and 3.0 km/sec for transit and rendezvous maneuvers) and 3.0 km on the return trip. Since, in the range of feasible exhaust velocities, LEO start mass always decreases with increasing exhaust velocity, a value of 8.0 km/s was selected and is used henceforth. For this value, the captured mass is 0.535×10^6 tons.

The sensitivity of LEO start mass and capture—the start mass ratio—to variations in outbound and return velocity requirements is shown in Figure 13. Note that the post-escape outbound velocity requirement doubles, but there is no more than about a

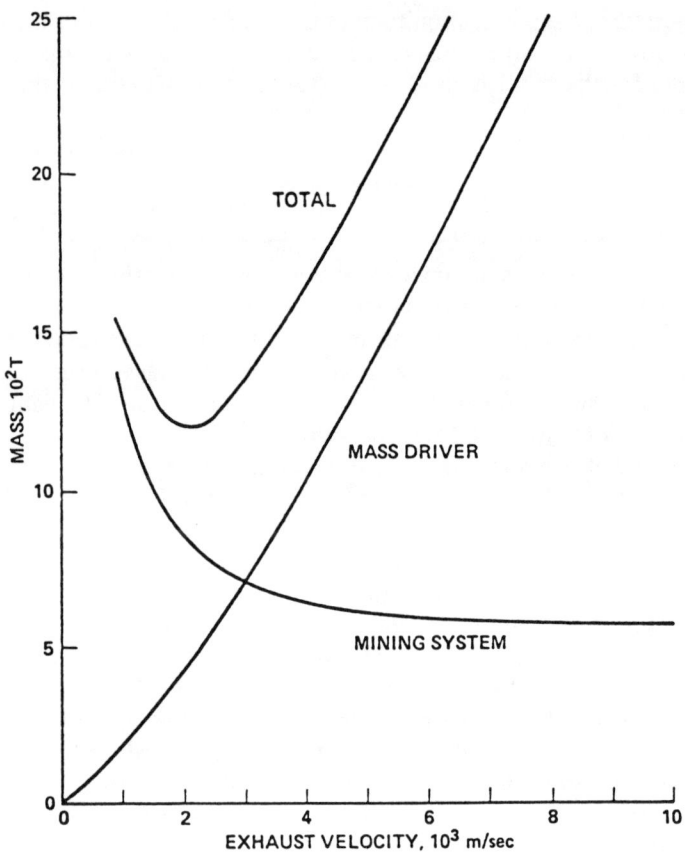

FIGURE 11.   Effects of exhaust velocity on asteroid mining vehicle mass.

45% increase in LEO start mass and a corresponding decrease in capture mass. Also, a 67% increase in return velocity requirement produces only a 21% reduction of capture start mass ratio. Therefore, it can be concluded that a comfortable range of potential asteroid orbits can be accommodated without severe losses.

Figure 14A is a schematic of the asteroid capture mission; it also includes a parametric cost equation expressing total program costs per kilogram of asteroid captured. Also shown for comparison is a similar expression for the corresponding costs of terrestrial material. In the terrestrial case, it is postulated that the material is procured on Earth, transported to LEO, and then carried to geosynchronous orbit by the same mass driver as used in the asteroid case, which subsequently returns itself to LEO to pick up another cargo. In the terrestrial case, costs are amortized over a cumulative delivered mass equal to the asteroid capture mass, processed to the same degree. (Notation for the equations in Figure 14A and numerical values assumed are summarized in Figure 14B.)

The cost expressions in Figure 14A are plotted in Figure 15 using the values assumed in Figure 14B. Six observations were made.

(1) For capture of a 10^6-ton asteroid, estimated total program costs are from \$12 to 14 billion or about \$24/kg of captured mass. If Research, Development, Test, and Engineering (RDT& E) are excluded, the captured mass cost reduces to about \$4/kg. These values assume up-rated Shuttle as the Earth-to-LEO transporter, with transportation costs of \$240/kg.

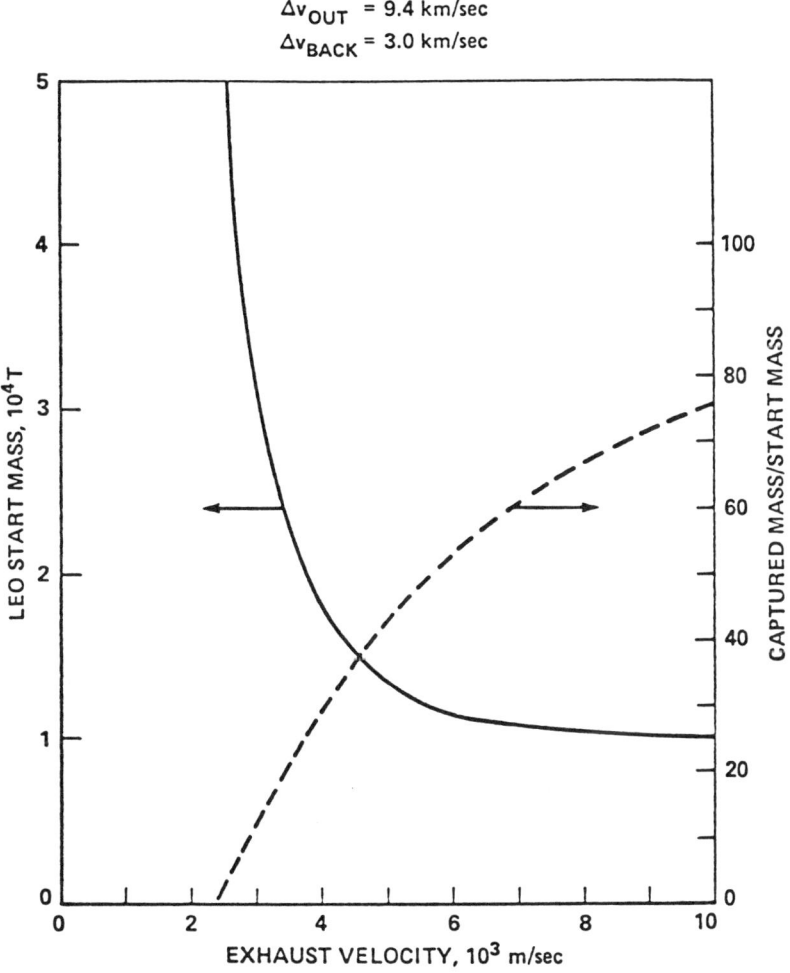

Δv_{OUT} = 9.4 km/sec
Δv_{BACK} = 3.0 km/sec

FIGURE 12.   Effects of exhaust velocity on start and capture mass.

(2) These results are only moderately sensitive to Earth-to-LEO transportation cost and reaction mass cost because of the relatively high RDT & E and hardware procurement costs.

(3) Delivery to geosynchronous orbit (GSO) of a total mass of terrestrial materials equal to the asteroid capture mass would involve, on the same basis of comparison, a total program cost of about $633 billion and a delivered cost/kilogram of about $355/kg, including RDT & E costs ($343/kg, excluding RDT & E costs).

(4) If the asteroid mass were increased, RDT & E and procurement costs would increase somewhat, but capture mass would increase proportionately with asteroid mass, resulting in significant reduction in cost/kg captured. Thus for a 10^6-ton asteroid, cost/kg captured might be reduced to about 50¢/kg.

(5) While support operations costs are not considered here, it seems likely that their magnitudes relative to the costs considered would be similar for both cases, and not dominant.

(6) This preliminary parametric assessment confirms suggestions that asteroids may be promising sources of materials for support of space activities on a growing scale in the future, and that this possibility should receive increased attention.

To estimate a common-based comparison of asteroidal and lunar materials costs,

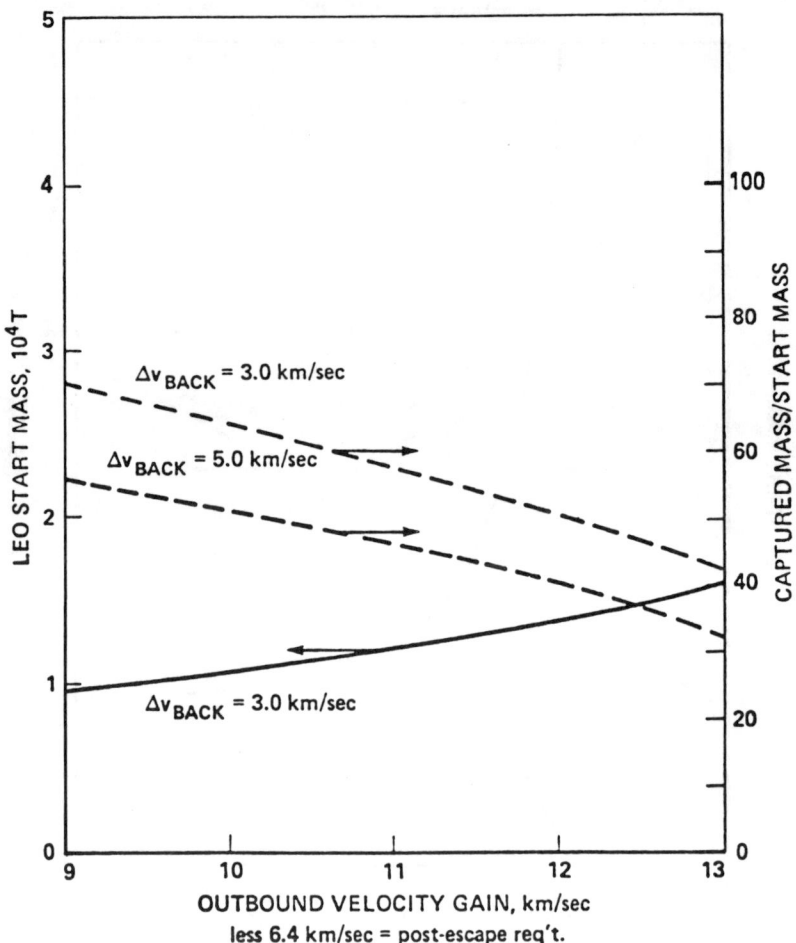

FIGURE 13. Effects of mission velocity on start and capture mass.

both are here placed on the same program basis as assumed for the lunar case.[5] That is, Earth launches are assumed to begin in 1985 using the first generation space shuttle (capacity—60 flights per year), phasing in 1987 to a shuttle-derivative heavy-lift vehicle (SD/HLV) for unmanned cargo (capacity—80 flights per year), and finally phasing in 1991 to a second generation passenger-cargo single-stage-to-orbit (SSTO) shuttle. Program duration is 10 years (1985-1994), during which about 1 million tons of asteroid material (from the capture of two 1-million-ton asteroids) are returned to the space manufacturing facility (SMF), compared with about 2.4 million tons of lunar materials.

Capture of two asteroids in this time period is based on the estimate of a 5-year out-and-back mission time, and is consistent with the maximum-paced program assumed for the lunar case. Thus, the first generation shuttle launches the asteroid miner and mass driver in sections during 1985-86, contributing in the process all of its external tanks for mass-driver reaction mass. The SD/HLV then phases in to launch a second miner and mass driver, and all the remaining reaction mass for two asteroid expeditions as well as chemical orbital transfer vehicles (OTV) and their propellants for establishing the SMF, during 1987-89. The first expedition leaves low-Earth orbit (LEO) early in 1987 and returns in 1992. When it has successfully reached its target and

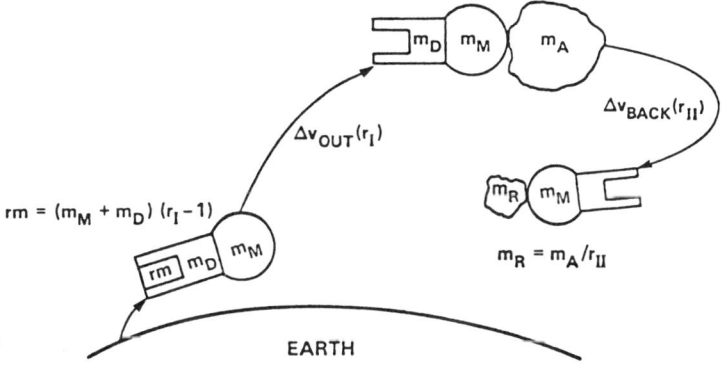

$$rm = (m_M + m_D) \, (r_I - 1)$$

$$m_R = m_A / r_{II}$$

EARTH

$$\text{MASS LIFTED TO LEO} = \frac{\text{HARDWARE}}{m_M + m_D} + \frac{\text{REACTION MASS}}{(m_M + m_D) \, (r_I - 1)}$$

$$\text{PROGRAM COST/kg CAPTURED} = \left[\frac{\text{MINING SYSTEM}}{m_M \, (Cd_M + Cp_M + C_t)} + \frac{\text{MASS DRIVER}}{m_D \, (Cd_D + Cp_D + C_T)} \right.$$

$$\left. + \frac{\text{REACTION MASS}}{(m_M + m_D) \, (r_I - 1) \, Crm} \right] \frac{r_{II}}{mA}$$

TERRESTRIAL MAT'L (REF.)

$$\text{PROGRAM COST/kg CAPTURED} = \left[\frac{\text{MASS DRIVER}}{m_D \, (Cd_D + Cp_D + C_t)} + \frac{\frac{m_A}{r_{II}} \, (r_{GSO} - 1) \, Crm}{\text{REACTION MASS}} \right.$$

$$\left. + \frac{\frac{m_A}{r_{II}} \, Cp_E}{\text{TERR. MAT'L.}} \right] \frac{r_{II}}{m_A}$$

FIGURE 14. Parametric schematic for asteroid retrieval.

is homeward bound in 1989, the second leaves to return in 1994. The SD/HLV, SSTO and OTV establish the SMF in 1989-92 so that it is ready to process the first returning asteroid. The chemical OTV is used for SMF rather than a smaller mass driver, to avoid a second mass driver development program, and because the OTV is require in any case for SMF personnel transfer. Use of a mass driver for deploying SMF would reduce total program cost only slightly.

The results of preliminary costing of the asteroid and lunar programs are summarized in Table 3. A possible requirement for an LEO station has been included as one consideration, to maintain as comparable basis as possible with the lunar case. Several observations may be made:

1. Costs of asteroidal materials are higher in Table 3 than in Figure 15 because of inclusion of the SMF and higher cost launch vehicles in the program, to be more comparable with the lunar case.
2. Costs of asteroidal materials are not sensitive to the requirement for an LEO station, since its development, deployment, and resupply costs are small in the total picture.
3. The total program cost for the asteroid option is only about half that of the lunar case including nonrecurring RDT & E costs; it is only about ⅓ that of the lunar case if RDT & E costs are excluded.

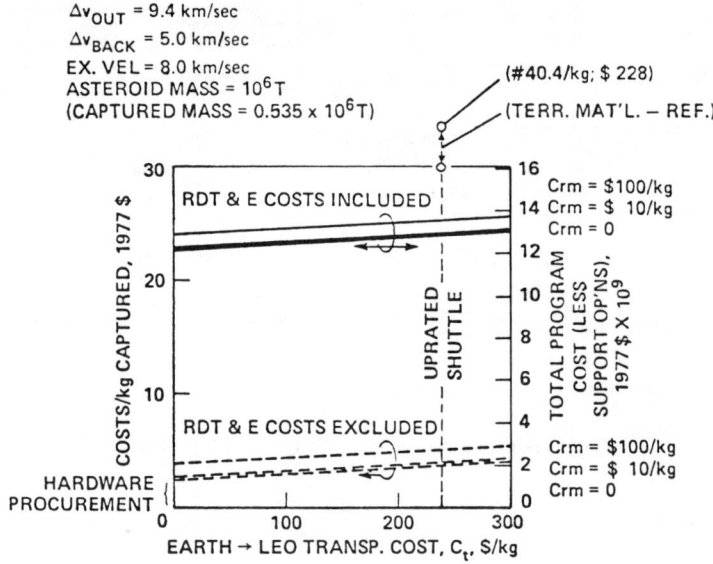

FIGURE 15.   Cost of captured asteroid mass.

Table 3
CONVENIENT MISSIONS TO TWO AMOR ASTERIODS

			Velocity intervals (ΔV) (km/sec)		
Object	Launch date	Escape from Leo[a]	Rendezvous with the Asteroid	Departure from the Asteroid	Capture in SMF Orbit[b]
433 Eros[19]	1993	4.7	1.8	1.7	0.3
1943 Anteros[20]	1992	4.6	0.9	1.6	0.3

[a] Assuming no lunar gravity assist on outbound leg.
[b] With two lunar gravity assists and a capture maneuver of 300 m/sec, where the excess hyperbolic velocity $v_\infty \leq 2.2$ km/sec.

4. Cost per delivered kg of material is 10 to 20% higher for asteroid materials if RDT & E are included, but 25 to 30 % lower than lunar materials on a recurring cost basis. Reasons for this are the relatively high RDT & E costs, but the lower amount of mass-driver throughput for the same delivered mass, for the asteroid than the lunar case. In addition, the asteroid operation is not burdened with the establishment and operation of anything akin to the mass-catcher or lunar orbit facilities and logistics.

5. These results suggest that for comparable or better economics in terms of cost per kilogram delivered, the asteroid option is significantly more attractive than the lunar option in terms of front-end "price of admission." The relative attractiveness of the asteroid option would of course be further strengthened is asteroidal material compositions are found more favorable than those of lunar materials.

VI. RECOMMENDATIONS

The 1977 Ames summer study has determined that, through the techniques of multiple gravity assists by the Earth, moon, and Venus, the total one-way Δv from Earth escape (or capture) to rendezvous (depart from) existing asteroids, in favorable cases, is from 2 to 4 km/sec. The mission analysis described here indicates that a single-stage, low-Earth-orbit to asteroid mass driver with an exhaust velocity of about 8 km/sec may be the most cost-effective alternative. Minimizing Δv in the outbound leg of the mission was found to be more important than minimizing inbound Δv, in terms of the amount of Earth-launched mass required to perform the retrieval mission. A large Δv for the inbound leg (e.g., 5 to 8 km/sec) increases the mass of the mining operation to prepare reaction mass, thereby reducing the amount of asteroidal material retrievable, but has no significant bearing on the total mass required to start the mission in low Earth orbit, which appears to be the major cost driver. But a modest increase in the Δv of the outbound leg to values up to 6 km/sec from escape would not impose severe penalties. Therefore, there are mission opportunities to existing targets that can be immediately investigated. Further refinements of mission-analysis techniques for known objects will permit a more precise determination of known opportunities. In addition, opportunities will arise as the number of known objects increases appreciably over the coming years.

The total asteroid-retrieving mass consists of the mass driver, mining equipment, and reaction mass for the boost. In the range of Δv values likely to be encountered, the total mass required for launch to low-Earth orbit to retrieve about half of 10^6-ton asteroid is near 10,000 tons, similar to the lunar case. During the mid to late 1980s, at a cost of approximately $2 billion, this magnitude of mass could be launched by an upgraded (class II) Shuttle with the external tanks used as reaction mass. Development costs (about $10 billion) would dominate and would be absorbed in subsequent, large-scale asteroidal retrievals. Alternatively, reaction mass could become available in low-Earth orbit in the form of lunar materials obtained in an early, modest progam of lunar mining.[6]

Perhaps the most unique aspect of retrieving asteroidal materials is the availability of large quantities of volatiles (water and carbon compounds) and free metals. These materials may not be available in large quantities on the moon. Since about 30,000 tons of consumables are required for space settlements that support the construction of satellite power stations, most of these consumables preferably could come from the asteroids rather than from Earth. Moreover, free metals separated at the asteroid and alloyed at the SMF into useful structures may eliminate many of the complex chemical processing steps required in the lunar case.

Of the 50,000 tons of asteroidal material that would arrive at the SMF, much of this could be water and carbon if a type I or II carbonaceous object at a low to moderate Δv could be discovered. It is possible that the Earth-approaching asteroid Betulia is carbonaceous, and statistical considerations suggest that such a discovery, in an expanded search and followup program, is probable at some time over the next few years.[12] Such an object would obviously be a prime target for a precursor mission.

The retrieval of asteroidal materials for space manufacturing appears to be cost-competitive and contemporary with the retrieval of lunar materials, and considerably less expensive than transporting consumables from the Earth if a type I or II carbonaceous object with reasonably favorable Δv characteristics could be found. Therefore, the asteroid option for space manufacturing should be kept open. The 1977 Ames summer study group[10] recommended that the following studies and programs be carried out immediately to better assess the details of this option.

Continued studies of retrieval mission opportunities: late in this study, we recognized that gravity assists in a low-thrust mission significantly reduce the mass requirements and that the Δv in the outbound leg should be minimized; also the economics of asteroid retrieval are not severely affected for moderate Δ values (one way, about 6 km/sec). From an analysis of the asteroid 1977 HB, the inbound Δv was found in one case to be from 2 to 3 km/sec. Many more cases—and a scheme for selecting the best ones—must be studied. Several known candidate asteroids can be studied over the next few years, and new ones as they are discovered.

Increased asteroid search and followup program: the current inventory of 39 Earth-approaching asteroids whose orbits are known can be increased appreciably with a modest investment in dedicated Earth-based telescopes and perhaps orbital telescopes. The followup work of orbital determination and chemical analysis is also important. Such an inventory would identify carbonaceous objects with low-to-moderate Δv opportunities.

Precursor missions: because of the potential importance of obtaining volatiles—and possibly free metals—from asteroids, clearly, a precursor mission to the prime carbonaceous candidate(s) is paramount. Such a mission is also important to assess mineralogical structure for designing the mining and processing apparatus. The 1977 Summer Group recommended a new start in the NASA budget aimed toward single- or multiple-asteroid rendezvous and landing missions that would answer these questions for one or more objects between 1982 and 1985.

Continuing studies of the potentials of asteroidal retrieval: these studies would continue to assess the tradeoffs between the retrieval of asteroidal, lunar, and Earth materials for space manufacturing in the light of new information as it comes in.

Technology support: design concepts of asteroid-retriever mass drivers and mining and processing equipment must be further developed; technology development milestones must be established in parallel with the lunar option.

The essence of these recommendations was corroborated by a number of scientists who participated in a later NASA workshop held in La Jolla, Calif.[9]

VII. RECENT ASTEROID RESEARCH[13]

A. New Observations

Helin, Shoemaker, Kowal, and co-workers have reported the discovery of several new Apollo and Amor asteroids, bringing the total to 47. Transfer orbits to some of these new objects are susceptible to low Δv missions.[14] Photometric observations indicate that one of these asteroids, Ra-Shalom, is a "C class" object, joining Betulia and Phobos as probable carbonaceous asteroids. If some of the carbonaceous meteorites are in fact derived from carbonaceous Apollo and Amor asteroids, it is likely that they contain a significant abundance of water (up to 20% by weight) and carbon. The retrieval of asteroidal water is of obvious importance to the conduct of a cost-effective program of space industrialization.

Broadband and narrowband photometry between 3 and 4 μm in the near-infrared of the main belt asteroid 1 Ceres have revealed an absorption feature centered near 3.0 μm, which is the first evidence for water of hydration in the surface material of an asteroid.[15] Lebofsky[28] is planning an observing program where newly discovered and recurring Apollo and Amor objects down to 14th visual magnitude can be tested for the presence of water. Apparitions of many of these objects occur at this magnitude, or brighter, and this will therefore be a significant contribution to asteroid prospecting research.

The search for asteroids near the Earth-Sun L_4 and L_5 Lagrange points has so far yielded negative results, but the difficulty of observation and large region of search conspire to select against discovery of all but the largest objects.[16] Any successful finding would be significant because of the low Δv of retrieval.

Although the 1977 summer study groups made strong recommendations to augument the Apollo/Amor asteroid search program to a dedicated large-aperture Schmidt,

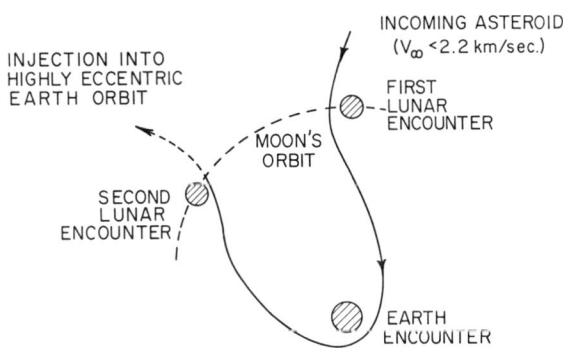

FIGURE 16. Geometry of Earth-capture of an asteroid with hyperbolic excess velocity, $V_x < 2.2$ km/sec and low inclination.

limitations in NASA funding for new programs have postponed these plans indefinitely. Meanwhile, observers will have to continue to rely on their own resources and resourcefulness.

B. Precursor Missions

The 1977 summer study groups also recommended a fiscal year 1980 new start on a program of rendezvous missions to selected Apollo and Amor asteroids for combined scientific and prospecting reasons. The recommendations were not followed by NASA, which has opted for a main belt multiple asteroid rendezvous mission utilizing ion propulsion as a first priority asteroid mission.[17] An Apollo or Amor mission is not in NASA's current planning for either the 1980s or 1990s even though some of these asteroids are the most accessible objects in the solar system in terms of energy.

C. Retrieval Mission Options

In spite of diminished NASA support of asteroid prospecting and retrieval concepts, a number of innovative concepts have been explored to minimize the energy cost of asteroid retrieval.

Ross and O'Leary have investigated the possibility of using two lunar gravity assists and one Earth encounter (miss distance about 1 Earth radius) to add as much as 2.2 km/sec to the escape velocity from the Earth on any interplanetary trajectory and, conversely, to "kill" as much as 2.2 km/sec excess hyperbolic velocity (V_x) to capture a payload in a highly elliptical orbit about Earth (Figure 16).

Alternatively, it is possible, by lunar gravity assists, to change the inclination of an interplanetary transfer orbit by as much as 5°. Closer encounters could raise these figures by some tens of percent while lowering the margin of safety. Transfer to (from) these highly elliptical Earth-capture orbits from (to) low-eccentricity (high Earth) orbits appropriate for space manufacturing require an equivalent ballistic thrust at perigee of about 300 m/sec. This figure can probably be reduced by a third lunar encounter-plus-thrusting over a period of several months.

Ross[18] has also shown that periodically applied low-thrust interplanetary trajectories for low Δv transfers (≤ 6 km/sec) were found to require the same Δv's as short-impulse ballistic transfers, i.e., the ballistic approximation appears to be a reasonable one for treating low-thrust missions.

To understand the consequences of these findings, we examine missions to the Amor asteroids 433 Eros and 1943 Anteros.[19,20] The return from 1943 Anteros is depicted in

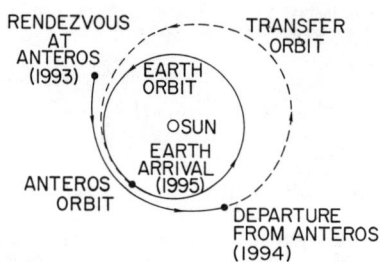

FIGURE 17.   Return from Anteros,
1993 to 1995.

Figure 17. The ballistic Δv's for favorable 1992 and 1993 launch dates are shown in Table 3.

If the mission objective is to retrieve tonnages of asteroidal material for high-orbital manufacturing, we see that in both cases the total return Δv is approximately 2 km/sec for zero inclination with respect to the Earth-moon system, suggesting the usage of low-specific-impulse and light weight rockets for the mission. Further analysis of mission opportunities may reveal more favorable opportunities in which asteroid departure Δv's are ≤ 1 km/sec (as a mirror image of the 0.9 km/sec Δv for rendezvous with 1943 Anteros) and where the total Δv for return approaches 1 km/sec. Such a finding could significantly reduce the unit cost of materials transport.

So far, I have discussed the direct return of materials from the asteroid to the Earth-moon system. Three other classes of encounters provide more options and further reduce the Δv requirements: planetary gravity assist, two-asteroid collision, and atmospheric graze orbit insertion. None of these methods constrains us to a direct low-inclination, $V_\infty < 2.2$ km/sec approach from the asteroid to the Earth-moon system. Bender et al.[11] investigated the return of the Apollo object 1977 HB (Figure 18) by means of one Earth, one Venus, and one or two additional Earth gravity assists (Figure 19). A 5-year return mission starting in 1984 (too soon to be realistic in terms of a materials return mission, but illustrative of what is possible) results in a total return Δv for thrusting of 3 km/sec.

An alternative plan, suggested by C. Singer,[21] is to use a column of debris of one asteroid to redirect another asteroid toward the Earth (or Venus). The rationale behind this collisional velocity change is that it is likely one can find a pair of asteroids in which the Δv required to redirect it toward another, and thence toward a planet, could be very low. With n being the number of known Apollo/Amor asteroids, there are n² possible candidate pairs. Alternatively, the number is only 2n in the case of planetary gravity assists where Venus and Earth are the only realistic candidates. With an increased search program, n ≥ 200, so that the collisional method could be quite attractive as a means for economic asteroid retrieval.

A third method of asteroid retrieval involves simply redirecting the velocity vector on a transfer orbit toward a grazing reentry corridor path toward Earth, where the material becomes injected into an appropriate Earth orbit for processing. We have seen the case where 1943 Anteros could be returned to Earth with an initial Δv of about 1 km/sec. With this technique, no collisions or planetary gravity assists are required, but care must be taken for accurate guidance. This would be possible if the asteroid were fragmented and optical scanning and electrostatic charging were carried out upon approach to the Earth, analogous to the fine guidance techniques considered for lunar material.

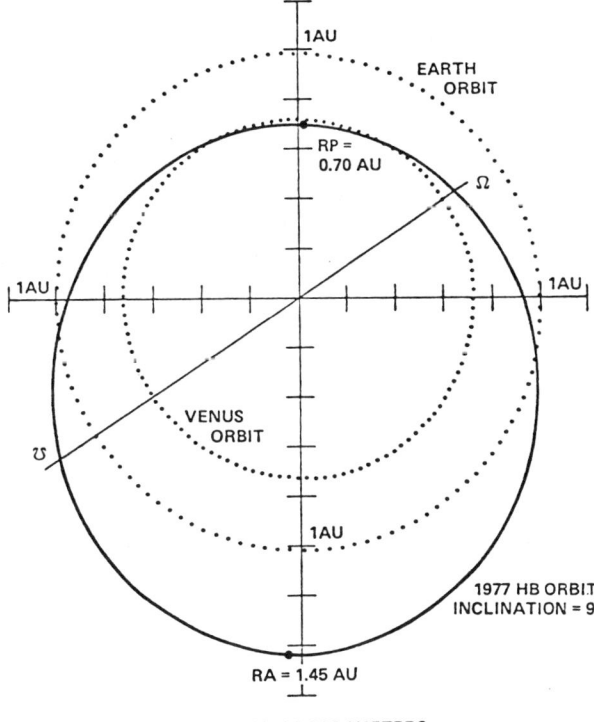

FIGURE 18. Orbit of 1977 HB.

PLANETARY GRAVITY ASSISTS

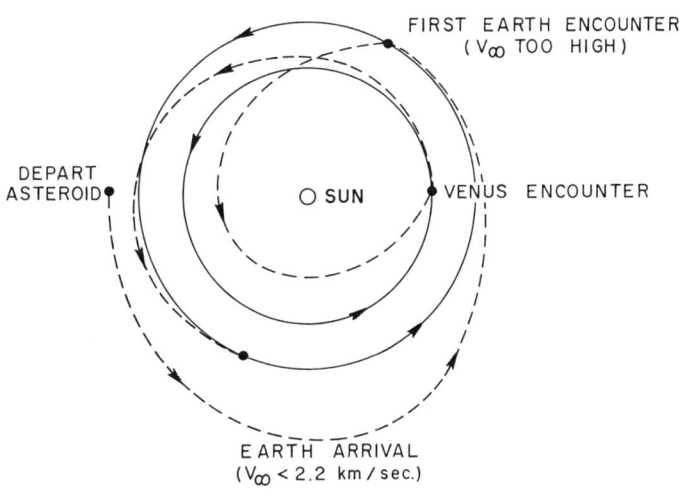

FIGURE 19. Planetary gravity assists.

D. Phobos as a Resource

Following a suggestion by S. F. Singer, von Herzen, a Princeton student, and O'Leary investigated the accessibility of the moon of Mars Phobos as a potential resource for space manufacturing. By utilizing a Martian atmospheric graze as a tool for braking the retriever spacecraft into an orbit about Mars in which Phobos is at apoapsis, the following ballistic Δv's were obtained: escape from LEO, 3.4 km/sec; rendezvous with Phobos, 0.5 km/sec; departure from Phobos on a Hohmann transfer, 1.9 km/sec; and some uncertain (but similar) value for capture into high Earth orbit (see Table 3).

Comparing these values to those in Table 3, we see that Phobos is currently the most accessible known natural object in the solar system. Moreover, Viking photographs showing a low albedo for Phobos suggest that the material comprising its surface is likely to be carbonaceous, with large quantities of water, carbon, and nitrogen available. The potential exists for a vast resource; a piece of Phobos the size of the smallest resolution element of a Viking photograph filled with Phobos is adequate to build several satellite power stations. Reasonably favorable launch windows to Phobos occur every 2 years, which are considerably more frequent than those for the Earth-approaching asteroids whose orbits are generally eccentric (for example, reasonable launch windows to 1943 Anteros and 433 Eros occur every 15 to 17 years).

Finally, a commitment to using the resources of Phobos would allow for setting up a base on Phobos for the exploration of Mars.

E. Other Rationales for Asteroid Prospecting and Utilization

Using meteorite data, Kuck[22] has suggested that certain classes of asteroids could provide rare metals which would be worth several billion dollars in terrestrial marketplaces. Gaffey and McCord[22] have analyzed the large market for nickel and iron which could accrue from the recovery on Earth of metal-rich asteroids. The author looked at the same possibility for growing food in space using asteroidal materials both for agricultural structures and as soil.[24] Taylor[25] has recently suggested that research on asteroid orbits and retrieval missions should be augmented because of a potentially catastrophic collision of a small asteroid with Earth which, although improbable during a given period of a few years, is still of sufficient concern to warrant consideration of preventative action (i.e., diverting the orbit of the asteroid.)

VIII. CONCLUSIONS

Work since the 1977 Ames Summer Study shows improved economies of Earth-approaching asteroid missions.[26] The Summer Study, which made more conservative assumptions about the asteroid case, showed that the economy of asteroid-materials retrieval appears to be more favorable than the lunar case. Also assumed in the Summer Study scenario was a massive, high-energy mass driver with exhaust velocities as much as 10 times greater (or 100 times the energy) than what may really be required for retrieval operations.

The asteroid case should be reexamined from an engineering and economics point of view in light of these findings. Rocket systems with low-specific impulse and either high or low thrust are candidates for the retrieval portion of the mission (e.g., low-energy mass drivers and chemical fuels, where fuels come from the asteroid itself).

It is unfortunate that support for these studies by NASA has been stopped. Moreover, augmented Earth-based search and followup activities and a program of precursor missions will need to be carried out prior to retrieval missions. As it will take several years for these programs to be completed, it is important to consider beginning them soon.

Asteroidal resources may become the most cost-effective basis for a large-scale program of space industrialization. The possibilities of providing abundant consumables to space inhabitants, of growing food in water-rich asteroid soil and directly recovering asteroidal metals on Earth provide incentives beyond satellite solar power for utilizing asteroidal resources.

REFERENCES

1. **Preston-Thomas, H.,** *J. Br. Interplanet. Soc.,* 11, 173, 1952; Cole, D. M. and Cox, D. W., *Islands in Space: The Challenge of the Planetoids,* Chilton, Radnor, Pa., 1964; Arrhenius, G., *Ambio,* 3, 130, 1974; Gaffey, M. J. and McCord, *Technol. Rev.,* 79, 50, 1977; O'Neil, G. K., in *Space Manufacturing Facilities* (proc. 1974, 1975 Princeton Conferences), American Institute of Aeronautics and Astronautics, Grey, J., Ed., New York, 1977; Drexler, E., in *Space Manufacturing Facilities,* (Proc. 1974, 1975 Princeton Conferences, American Institute of Aeronautics and Astronautics, Grey, J., Ed., New York, 1977; Johnson, R. D. and Holbrow, C., Eds., Space Settlements: A Design Study, NASA SP-413, National Aeronautics and Space Administration, Washington, 1977; O'Neill, G. K., *The High Frontier: Human Colonies in Space,* Morrow, New York, 1977.
2. **O'Neil, G. K.,** The colonization of space, *Phys. Today,* 27, 32, 1974.
3. **O'Neill, G. K.,** Space colonies and energy supply to the Earth, *Science,* 190, 943, 1975.
4. **O'Neill, G. K., and O'Leary, B. T., Eds.,** *Space Manufacturing from Nonterrestrial Materials,* Vol. 57, American Institute of Aeronautics and Astronautics, New York, 1978.
5. **Billingham, J., Gilbreath, W., and O'Leary, B., Eds.,** Space Resources and Space Settlements, (NASA SP-428), National Aeronautics and Space Administration, Washington, D.C., 1979.
6. **O'Neill, G. K.,** The low (profile) road to space manufacturing, *Astronaut. Aeronaut.,* 16, 24, 1978.
7. **O'Leary, B.,** Mining the Apollo and Amor asteroids, *Science,* 197, 363, 1977.
8. **O'Leary, B.,** Mass driver retrievals of Earth-approaching asteroids, in *Space Manufacturing Facilities II,* Grey, J., Ed., American Institute of Aeronautics and Astronautics, New York, 1977, 157.
9. **Arnold, J. R.,** Ed., *Summer Worshop on Near-Earth Resources,* La Jolla, Calif. August 12, 1977, NASA Conference Publication 2031, Washington, D.C., 1977.
10. **O'Leary, B., Gaffey, M. J., Ross, D., and Salkeld, R.,** The retrieval of asteroidal materials, in *Space Resour. Space Settlements,* Billingham, J., Gilbreath, W., and O'Leary, B., Eds., NASA SP-428, National Aeronautics and Space Administration, Washington, D.C., 1979.
11. **Bender, D. F., Dunbar, R. S., and Ross, D. J.,** Round-trip missions to low-delta-v asteroids and implications for material retrieval, in *Space Resour. Space Settlements,* Billingham, J., Gilbreath, W., and O'Leary, B., Eds., NASA SP-428, National Aeronautics and Space Administration, Washington, D.C., 1979.
12. **Gaffey, M. J., Helin, E. F., and O'Leary, B.,** An assessment of near-Earth asteroid resources, in *Space Resour. Space Settlements,* Billingham, J., Gilbreath, W., and O'Leary, B., Eds., NASA SP-428, National Aeronautics and Space Administration, Washington, D.C., 1979.
13. **O'Leary, B.,** Asteroid prospecting and retrieval, in *Space Manufacturing Facilities III,* Grey, J., Ed., American Institute of Aeronautics and Astronautics, New York, 1980.
14. **Shoemaker, E.,** *The Arizona Conference on Asteroids,* March 6—11, 1979 University of Arizona Press, Tucson, Ariz., 1980.
15. **Lebofsky, L.,** *Monthly Notes R. Astron. Soc.,* 182, 17, 1978.
16. **Dunbar, R. S.,** The search for asteroids in the L_4 and L_5 libration points in the Earth-Sun system, in *Space Manufacturing Facilities III,* Grey, J., Ed., American Institute of Aeronautics and Astronautics, New York (in press).
17. **Morrison, D. and Wells, W.,** Eds., *Asteroids: An Exploration Assessment,* Workshop held at the University of Chicago, January 19-21, 1978, NASA Conference Publication 2053, Washington, D.C., 1978.
18. **Ross, D.,** Low-thrust alteration of asteroidal orbits, in *Space Manufacturing Facilities III,* Grey, J. Ed., American Institute of Aeronautics and Astronautics, New York 1980.
19. **Mascy, A. C. and Niehoff, J.,** in *Phys. Studies of Minor Planets,* Gehrels, T., Ed., NASA SP-267, National Aeronautics and Space Administration, Washington, D.C., 1971, 473.
20. **Niehoff, J.,** *Icarus,* 31, 430, 1977.
21. **Singer, C.,** Collisional orbital change of asteroidal materials, in *Space Manuf. Facilities III,* Grey, J., Ed., American Institute of Aeronautics and Astronautics, New York, 1980.
22. **Kuck, D. L.,** Near-Earth extraterrestrial resources, in *Space Manuf. Facilities III,* Grey J., Ed., American Institue of Aeronautics and Astronautics, New York, 1980.
23. **Gaffey, M. J., and McCord, T. C.,** Mining outer space, *Technol. Rev.,* 79, 50, 1977.
24. **O'Leary, B.,** Food and raw material supply from space to the Earth, in *Escasez Mundial de Alimentos y Materias Primas,* Vicuña, F. O., Ed., Universidad de Chile, Santiago, 1977; Limits to growth implications of space settlement, presented at AAAS Symp. Prospects for Life in the Universe: The Ultimate Limits to Growth, Washington, D.C. February 12 to 17, 1978; Space agriculture and the world food problem, *Space Humanization Series,* 1, 115, 1979.
25. **Taylor, T.,** Effects of large object collisions with the Earth, in *Space Manu. Facilities III,* Grey, J., Ed., American Institute of Aeronautics and Astronautics, New York (in press).

26. **Salkeld, R.,** Parametric analysis of the comparative cost of recovering terrestrial, lunar and asteroidal materials, in *Space Manuf. Facilities III,* Grey, J., Ed., American Institute of Aeronautics and Astronautics, New York, 1980.

27. **Bock, E.,** Convair Division, Personal communication, 1977.

28. **Lebofsky, L.,** Private communication, 1979.

INDEX